Crystallography: A Very Short Introduction

VERY SHORT INTRODUCTIONS are for anyone wanting a stimulating and accessible way into a new subject. They are written by experts, and have been translated into more than 40 different languages.

The series began in 1995, and now covers a wide variety of topics in every discipline. The VSI library now contains over 450 volumes—a Very Short Introduction to everything from Psychology and Philosophy of Science to American History and Relativity—and continues to grow in every subject area.

Very Short Introductions available now:

ACCOUNTING Christopher Nobes
ADVERTISING Winston Fletcher
AFRICAN AMERICAN RELIGION
    Eddie S. Glaude Jr
AFRICAN HISTORY John Parker and
    Richard Rathbone
AFRICAN RELIGIONS Jacob K. Olupona
AGNOSTICISM Robin Le Poidevin
ALEXANDER THE GREAT Hugh Bowden
ALGEBRA Peter M. Higgins
AMERICAN HISTORY Paul S. Boyer
AMERICAN IMMIGRATION
    David A. Gerber
AMERICAN LEGAL HISTORY
    G. Edward White
AMERICAN POLITICAL HISTORY
    Donald Critchlow
AMERICAN POLITICAL PARTIES
    AND ELECTIONS L. Sandy Maisel
AMERICAN POLITICS Richard M. Valelly
THE AMERICAN PRESIDENCY
    Charles O. Jones
THE AMERICAN REVOLUTION
    Robert J. Allison
AMERICAN SLAVERY
    Heather Andrea Williams
THE AMERICAN WEST Stephen Aron
AMERICAN WOMEN'S HISTORY
    Susan Ware
ANAESTHESIA Aidan O'Donnell
ANARCHISM Colin Ward
ANCIENT ASSYRIA Karen Radner
ANCIENT EGYPT Ian Shaw
ANCIENT EGYPTIAN ART AND
    ARCHITECTURE Christina Riggs
ANCIENT GREECE Paul Cartledge

THE ANCIENT NEAR EAST
    Amanda H. Podany
ANCIENT PHILOSOPHY Julia Annas
ANCIENT WARFARE Harry Sidebottom
ANGELS David Albert Jones
ANGLICANISM Mark Chapman
THE ANGLO-SAXON AGE John Blair
THE ANIMAL KINGDOM
    Peter Holland
ANIMAL RIGHTS David DeGrazia
THE ANTARCTIC Klaus Dodds
ANTISEMITISM Steven Beller
ANXIETY Daniel Freeman and
    Jason Freeman
THE APOCRYPHAL GOSPELS
    Paul Foster
ARCHAEOLOGY Paul Bahn
ARCHITECTURE Andrew Ballantyne
ARISTOCRACY William Doyle
ARISTOTLE Jonathan Barnes
ART HISTORY Dana Arnold
ART THEORY Cynthia Freeland
ASTROBIOLOGY David C. Catling
ATHEISM Julian Baggini
AUGUSTINE Henry Chadwick
AUSTRALIA Kenneth Morgan
AUTISM Uta Frith
THE AVANT GARDE David Cottington
THE AZTECS David Carrasco
BACTERIA Sebastian G. B. Amyes
BARTHES Jonathan Culler
THE BEATS David Sterritt
BEAUTY Roger Scruton
BESTSELLERS John Sutherland
THE BIBLE John Riches
BIBLICAL ARCHAEOLOGY Eric H. Cline

BIOGRAPHY Hermione Lee
BLACK HOLES Katherine Blundell
THE BLUES Elijah Wald
THE BODY Chris Shilling
THE BOOK OF MORMON Terryl Givens
BORDERS Alexander C. Diener and
    Joshua Hagen
THE BRAIN Michael O'Shea
THE BRITISH CONSTITUTION
    Martin Loughlin
THE BRITISH EMPIRE Ashley Jackson
BRITISH POLITICS Anthony Wright
BUDDHA Michael Carrithers
BUDDHISM Damien Keown
BUDDHIST ETHICS Damien Keown
BYZANTIUM Peter Sarris
CANCER Nicholas James
CAPITALISM James Fulcher
CATHOLICISM Gerald O'Collins
CAUSATION Stephen Mumford and
    Rani Lill Anjum
THE CELL Terence Allen and
    Graham Cowling
THE CELTS Barry Cunliffe
CHAOS Leonard Smith
CHEMISTRY Peter Atkins
CHILD PSYCHOLOGY Usha Goswami
CHILDREN'S LITERATURE
    Kimberley Reynolds
CHINESE LITERATURE Sabina Knight
CHOICE THEORY Michael Allingham
CHRISTIAN ART Beth Williamson
CHRISTIAN ETHICS D. Stephen Long
CHRISTIANITY Linda Woodhead
CITIZENSHIP Richard Bellamy
CIVIL ENGINEERING David Muir Wood
CLASSICAL LITERATURE William Allan
CLASSICAL MYTHOLOGY
    Helen Morales
CLASSICS Mary Beard and
    John Henderson
CLAUSEWITZ Michael Howard
CLIMATE Mark Maslin
CLIMATE CHANGE Mark Maslin
THE COLD WAR Robert McMahon
COLONIAL AMERICA Alan Taylor
COLONIAL LATIN AMERICAN
    LITERATURE Rolena Adorno
COMEDY Matthew Bevis
COMMUNISM Leslie Holmes
COMPLEXITY John H. Holland
THE COMPUTER Darrel Ince

COMPUTER SCIENCE Subrata Dasgupta
CONFUCIANISM Daniel K. Gardner
THE CONQUISTADORS Matthew Restall
    and Felipe Fernández-Armesto
CONSCIENCE Paul Strohm
CONSCIOUSNESS Susan Blackmore
CONTEMPORARY ART
    Julian Stallabrass
CONTEMPORARY FICTION
    Robert Eaglestone
CONTINENTAL PHILOSOPHY
    Simon Critchley
CORAL REEFS Charles Sheppard
CORPORATE SOCIAL
    RESPONSIBILITY Jeremy Moon
CORRUPTION Leslie Holmes
COSMOLOGY Peter Coles
CRIME FICTION Richard Bradford
CRIMINAL JUSTICE Julian V. Roberts
CRITICAL THEORY Stephen Eric Bronner
THE CRUSADES Christopher Tyerman
CRYPTOGRAPHY Fred Piper and
    Sean Murphy
CRYSTALLOGRAPHY A. M. Glazer
THE CULTURAL REVOLUTION
    Richard Curt Kraus
DADA AND SURREALISM
    David Hopkins
DANTE Peter Hainsworth and David Robey
DARWIN Jonathan Howard
THE DEAD SEA SCROLLS Timothy Lim
DEMOCRACY Bernard Crick
DERRIDA Simon Glendinning
DESCARTES Tom Sorell
DESERTS Nick Middleton
DESIGN John Heskett
DEVELOPMENTAL BIOLOGY
    Lewis Wolpert
THE DEVIL Darren Oldridge
DIASPORA Kevin Kenny
DICTIONARIES Lynda Mugglestone
DINOSAURS David Norman
DIPLOMACY Joseph M. Siracusa
DOCUMENTARY FILM
    Patricia Aufderheide
DREAMING J. Allan Hobson
DRUGS Leslie Iversen
DRUIDS Barry Cunliffe
EARLY MUSIC Thomas Forrest Kelly
THE EARTH Martin Redfern
EARTH SYSTEM SCIENCE Tim Lenton
ECONOMICS Partha Dasgupta

EDUCATION Gary Thomas
EGYPTIAN MYTH Geraldine Pinch
EIGHTEENTH-CENTURY BRITAIN
 Paul Langford
THE ELEMENTS Philip Ball
EMOTION Dylan Evans
EMPIRE Stephen Howe
ENGELS Terrell Carver
ENGINEERING David Blockley
ENGLISH LITERATURE Jonathan Bate
THE ENLIGHTENMENT John Robertson
ENTREPRENEURSHIP Paul Westhead
 and Mike Wright
ENVIRONMENTAL ECONOMICS
 Stephen Smith
ENVIRONMENTAL POLITICS
 Andrew Dobson
EPICUREANISM Catherine Wilson
EPIDEMIOLOGY Rodolfo Saracci
ETHICS Simon Blackburn
ETHNOMUSICOLOGY Timothy Rice
THE ETRUSCANS Christopher Smith
THE EUROPEAN UNION John Pinder
 and Simon Usherwood
EVOLUTION Brian and
 Deborah Charlesworth
EXISTENTIALISM Thomas Flynn
EXPLORATION Stewart A. Weaver
THE EYE Michael Land
FAMILY LAW Jonathan Herring
FASCISM Kevin Passmore
FASHION Rebecca Arnold
FEMINISM Margaret Walters
FILM Michael Wood
FILM MUSIC Kathryn Kalinak
THE FIRST WORLD WAR
 Michael Howard
FOLK MUSIC Mark Slobin
FOOD John Krebs
FORENSIC PSYCHOLOGY David Canter
FORENSIC SCIENCE Jim Fraser
FORESTS Jaboury Ghazoul
FOSSILS Keith Thomson
FOUCAULT Gary Gutting
THE FOUNDING FATHERS
 R. B. Bernstein
FRACTALS Kenneth Falconer
FREE SPEECH Nigel Warburton
FREE WILL Thomas Pink
FRENCH LITERATURE John D. Lyons
THE FRENCH REVOLUTION
 William Doyle

FREUD Anthony Storr
FUNDAMENTALISM Malise Ruthven
FUNGI Nicholas P. Money
GALAXIES John Gribbin
GALILEO Stillman Drake
GAME THEORY Ken Binmore
GANDHI Bhikhu Parekh
GENES Jonathan Slack
GENIUS Andrew Robinson
GEOGRAPHY John Matthews and
 David Herbert
GEOPOLITICS Klaus Dodds
GERMAN LITERATURE Nicholas Boyle
GERMAN PHILOSOPHY Andrew Bowie
GLOBAL CATASTROPHES Bill McGuire
GLOBAL ECONOMIC HISTORY
 Robert C. Allen
GLOBALIZATION Manfred Steger
GOD John Bowker
GOETHE Ritchie Robertson
THE GOTHIC Nick Groom
GOVERNANCE Mark Bevir
THE GREAT DEPRESSION AND THE
 NEW DEAL Eric Rauchway
HABERMAS James Gordon Finlayson
HAPPINESS Daniel M. Haybron
HEGEL Peter Singer
HEIDEGGER Michael Inwood
HERMENEUTICS Jens Zimmermann
HERODOTUS Jennifer T. Roberts
HIEROGLYPHS Penelope Wilson
HINDUISM Kim Knott
HISTORY John H. Arnold
THE HISTORY OF ASTRONOMY
 Michael Hoskin
THE HISTORY OF CHEMISTRY
 William H. Brock
THE HISTORY OF LIFE Michael Benton
THE HISTORY OF MATHEMATICS
 Jacqueline Stedall
THE HISTORY OF MEDICINE
 William Bynum
THE HISTORY OF TIME
 Leofranc Holford-Strevens
HIV/AIDS Alan Whiteside
HOBBES Richard Tuck
HOLLYWOOD Peter Decherney
HORMONES Martin Luck
HUMAN ANATOMY Leslie Klenerman
HUMAN EVOLUTION Bernard Wood
HUMAN RIGHTS Andrew Clapham
HUMANISM Stephen Law

HUME A. J. Ayer
HUMOUR Noël Carroll
THE ICE AGE Jamie Woodward
IDEOLOGY Michael Freeden
INDIAN PHILOSOPHY Sue Hamilton
INFECTIOUS DISEASE Marta L. Wayne
    and Benjamin M. Bolker
INFORMATION Luciano Floridi
INNOVATION Mark Dodgson and
    David Gann
INTELLIGENCE Ian J. Deary
INTERNATIONAL LAW Vaughan Lowe
INTERNATIONAL MIGRATION
    Khalid Koser
INTERNATIONAL RELATIONS
    Paul Wilkinson
INTERNATIONAL SECURITY
    Christopher S. Browning
IRAN Ali M. Ansari
ISLAM Malise Ruthven
ISLAMIC HISTORY Adam Silverstein
ITALIAN LITERATURE
    Peter Hainsworth and David Robey
JESUS Richard Bauckham
JOURNALISM Ian Hargreaves
JUDAISM Norman Solomon
JUNG Anthony Stevens
KABBALAH Joseph Dan
KAFKA Ritchie Robertson
KANT Roger Scruton
KEYNES Robert Skidelsky
KIERKEGAARD Patrick Gardiner
KNOWLEDGE Jennifer Nagel
THE KORAN Michael Cook
LANDSCAPE ARCHITECTURE
    Ian H. Thompson
LANDSCAPES AND
    GEOMORPHOLOGY
    Andrew Goudie and Heather Viles
LANGUAGES Stephen R. Anderson
LATE ANTIQUITY Gillian Clark
LAW Raymond Wacks
THE LAWS OF THERMODYNAMICS
    Peter Atkins
LEADERSHIP Keith Grint
LIBERALISM Michael Freeden
LIGHT Ian Walmsley
LINCOLN Allen C. Guelzo
LINGUISTICS Peter Matthews
LITERARY THEORY Jonathan Culler
LOCKE John Dunn
LOGIC Graham Priest

LOVE Ronald de Sousa
MACHIAVELLI Quentin Skinner
MADNESS Andrew Scull
MAGIC Owen Davies
MAGNA CARTA Nicholas Vincent
MAGNETISM Stephen Blundell
MALTHUS Donald Winch
MANAGEMENT John Hendry
MAO Delia Davin
MARINE BIOLOGY Philip V. Mladenov
THE MARQUIS DE SADE John Phillips
MARTIN LUTHER Scott H. Hendrix
MARTYRDOM Jolyon Mitchell
MARX Peter Singer
MATERIALS Christopher Hall
MATHEMATICS Timothy Gowers
THE MEANING OF LIFE Terry Eagleton
MEDICAL ETHICS Tony Hope
MEDICAL LAW Charles Foster
MEDIEVAL BRITAIN John Gillingham
    and Ralph A. Griffiths
MEDIEVAL LITERATURE Elaine Treharne
MEDIEVAL PHILOSOPHY
    John Marenbon
MEMORY Jonathan K. Foster
METAPHYSICS Stephen Mumford
THE MEXICAN REVOLUTION
    Alan Knight
MICHAEL FARADAY Frank A. J. L. James
MICROBIOLOGY Nicholas P. Money
MICROECONOMICS Avinash Dixit
MICROSCOPY Terence Allen
THE MIDDLE AGES Miri Rubin
MINERALS David Vaughan
MODERN ART David Cottington
MODERN CHINA Rana Mitter
MODERN DRAMA
    Kirsten E. Shepherd-Barr
MODERN FRANCE Vanessa R. Schwartz
MODERN IRELAND Senia Pašeta
MODERN JAPAN
    Christopher Goto-Jones
MODERN LATIN AMERICAN
    LITERATURE
    Roberto González Echevarría
MODERN WAR Richard English
MODERNISM Christopher Butler
MOLECULES Philip Ball
THE MONGOLS Morris Rossabi
MOONS David A. Rothery
MORMONISM Richard Lyman Bushman
MOUNTAINS Martin F. Price

MUHAMMAD Jonathan A. C. Brown
MULTICULTURALISM Ali Rattansi
MUSIC Nicholas Cook
MYTH Robert A. Segal
THE NAPOLEONIC WARS Mike Rapport
NATIONALISM Steven Grosby
NELSON MANDELA Elleke Boehmer
NEOLIBERALISM Manfred Steger
    and Ravi Roy
NETWORKS Guido Caldarelli and
    Michele Catanzaro
THE NEW TESTAMENT
    Luke Timothy Johnson
THE NEW TESTAMENT AS
    LITERATURE Kyle Keefer
NEWTON Robert Iliffe
NIETZSCHE Michael Tanner
NINETEENTH-CENTURY BRITAIN
    Christopher Harvie and
    H. C. G. Matthew
THE NORMAN CONQUEST
    George Garnett
NORTH AMERICAN INDIANS
    Theda Perdue and Michael D. Green
NORTHERN IRELAND Marc Mulholland
NOTHING Frank Close
NUCLEAR PHYSICS Frank Close
NUCLEAR POWER Maxwell Irvine
NUCLEAR WEAPONS
    Joseph M. Siracusa
NUMBERS Peter M. Higgins
NUTRITION David A. Bender
OBJECTIVITY Stephen Gaukroger
THE OLD TESTAMENT
    Michael D. Coogan
THE ORCHESTRA D. Kern Holoman
ORGANIZATIONS Mary Jo Hatch
PAGANISM Owen Davies
THE PALESTINIAN-ISRAELI
    CONFLICT Martin Bunton
PARTICLE PHYSICS Frank Close
PAUL E. P. Sanders
PEACE Oliver P. Richmond
PENTECOSTALISM William K. Kay
THE PERIODIC TABLE Eric R. Scerri
PHILOSOPHY Edward Craig
PHILOSOPHY IN THE ISLAMIC
    WORLD Peter Adamson
PHILOSOPHY OF LAW Raymond Wacks
PHILOSOPHY OF SCIENCE
    Samir Okasha
PHOTOGRAPHY Steve Edwards

PHYSICAL CHEMISTRY Peter Atkins
PILGRIMAGE Ian Reader
PLAGUE Paul Slack
PLANETS David A. Rothery
PLANTS Timothy Walker
PLATE TECTONICS Peter Molnar
PLATO Julia Annas
POLITICAL PHILOSOPHY David Miller
POLITICS Kenneth Minogue
POSTCOLONIALISM Robert Young
POSTMODERNISM Christopher Butler
POSTSTRUCTURALISM Catherine Belsey
PREHISTORY Chris Gosden
PRESOCRATIC PHILOSOPHY
    Catherine Osborne
PRIVACY Raymond Wacks
PROBABILITY John Haigh
PROGRESSIVISM Walter Nugent
PROTESTANTISM Mark A. Noll
PSYCHIATRY Tom Burns
PSYCHOANALYSIS Daniel Pick
PSYCHOLOGY Gillian Butler and
    Freda McManus
PSYCHOTHERAPY Tom Burns and
    Eva Burns-Lundgren
PURITANISM Francis J. Bremer
THE QUAKERS Pink Dandelion
QUANTUM THEORY John Polkinghorne
RACISM Ali Rattansi
RADIOACTIVITY Claudio Tuniz
RASTAFARI Ennis B. Edmonds
THE REAGAN REVOLUTION Gil Troy
REALITY Jan Westerhoff
THE REFORMATION Peter Marshall
RELATIVITY Russell Stannard
RELIGION IN AMERICA Timothy Beal
THE RENAISSANCE Jerry Brotton
RENAISSANCE ART Geraldine A. Johnson
REVOLUTIONS Jack A. Goldstone
RHETORIC Richard Toye
RISK Baruch Fischhoff and John Kadvany
RITUAL Barry Stephenson
RIVERS Nick Middleton
ROBOTICS Alan Winfield
ROMAN BRITAIN Peter Salway
THE ROMAN EMPIRE Christopher Kelly
THE ROMAN REPUBLIC
    David M. Gwynn
ROMANTICISM Michael Ferber
ROUSSEAU Robert Wokler
RUSSELL A. C. Grayling
RUSSIAN HISTORY Geoffrey Hosking

RUSSIAN LITERATURE Catriona Kelly
THE RUSSIAN REVOLUTION
    S. A. Smith
SCHIZOPHRENIA Chris Frith and
    Eve Johnstone
SCHOPENHAUER
    Christopher Janaway
SCIENCE AND RELIGION Thomas Dixon
SCIENCE FICTION David Seed
THE SCIENTIFIC REVOLUTION
    Lawrence M. Principe
SCOTLAND Rab Houston
SEXUALITY Véronique Mottier
SHAKESPEARE'S COMEDIES Bart van Es
SIKHISM Eleanor Nesbitt
THE SILK ROAD James A. Millward
SLANG Jonathon Green
SLEEP Steven W. Lockley and
    Russell G. Foster
SOCIAL AND CULTURAL
    ANTHROPOLOGY
    John Monaghan and Peter Just
SOCIAL PSYCHOLOGY Richard J. Crisp
SOCIAL WORK Sally Holland and
    Jonathan Scourfield
SOCIALISM Michael Newman
SOCIOLINGUISTICS John Edwards
SOCIOLOGY Steve Bruce
SOCRATES C. C. W. Taylor
SOUND Mike Goldsmith
THE SOVIET UNION Stephen Lovell
THE SPANISH CIVIL WAR Helen Graham
SPANISH LITERATURE Jo Labanyi
SPINOZA Roger Scruton
SPIRITUALITY Philip Sheldrake
SPORT Mike Cronin
STARS Andrew King
STATISTICS David J. Hand
STEM CELLS Jonathan Slack
STRUCTURAL ENGINEERING
    David Blockley

STUART BRITAIN John Morrill
SUPERCONDUCTIVITY
    Stephen Blundell
SYMMETRY Ian Stewart
TAXATION Stephen Smith
TEETH Peter S. Ungar
TERRORISM Charles Townshend
THEATRE Marvin Carlson
THEOLOGY David F. Ford
THOMAS AQUINAS Fergus Kerr
THOUGHT Tim Bayne
TIBETAN BUDDHISM
    Matthew T. Kapstein
TOCQUEVILLE Harvey C. Mansfield
TRAGEDY Adrian Poole
THE TROJAN WAR Eric H. Cline
TRUST Katherine Hawley
THE TUDORS John Guy
TWENTIETH-CENTURY
    BRITAIN Kenneth O. Morgan
THE UNITED NATIONS
    Jussi M. Hanhimäki
THE U.S. CONGRESS Donald A. Ritchie
THE U.S. SUPREME COURT
    Linda Greenhouse
UTOPIANISM Lyman Tower Sargent
THE VIKINGS Julian Richards
VIRUSES Dorothy H. Crawford
WATER John Finney
THE WELFARE STATE David Garland
WILLIAM SHAKESPEARE
    Stanley Wells
WITCHCRAFT Malcolm Gaskill
WITTGENSTEIN A. C. Grayling
WORK Stephen Fineman
WORLD MUSIC Philip Bohlman
THE WORLD TRADE
    ORGANIZATION Amrita Narlikar
WORLD WAR II Gerhard L. Weinberg
WRITING AND SCRIPT
    Andrew Robinson

Available soon:

ASTROPHYSICS James Binney
AGRICULTURE Paul Brassley and
    Richard Soffe

ISOTOPES Rob Ellam
BRICS Andrew F. Cooper
COMBINATORICS Robin Wilson

For more information visit our website

www.oup.com/vsi/

A. M. Glazer

CRYSTALLOGRAPHY

A Very Short Introduction

OXFORD
UNIVERSITY PRESS

OXFORD

UNIVERSITY PRESS

Great Clarendon Street, Oxford, OX2 6DP,
United Kingdom

Oxford University Press is a department of the University of Oxford.
It furthers the University's objective of excellence in research, scholarship,
and education by publishing worldwide. Oxford is a registered trade mark of
Oxford University Press in the UK and in certain other countries

© A.M. Glazer 2016

The moral rights of the author have been asserted

First edition published in 2016

Published in the United States of America by Oxford University Press
198 Madison Avenue, New York, NY 10016, United States of America

British Library Cataloguing in Publication Data
Data available

Library of Congress Control Number: 2015958968

ISBN 978-0-19-871759-1

Printed and bound by
CPI Group (UK) Ltd, Croydon, CR0 4YY

Links to third party websites are provided by Oxford in good faith and
for information only. Oxford disclaims any responsibility for the materials
contained in any third party website referenced in this work.

Contents

Preface xiii

List of illustrations xix

1 A long history 1

2 Symmetry 24

3 Crystal structures 39

4 Diffraction 64

5 Seeing atoms 94

6 Sources of radiation 107

Further reading 129

Index 133

Preface

Κρύσταλλος: rock crystal or ice (Greek)

'We also have an aura, made up of energy layers, vibrating at different frequencies. With healing crystals, chakra stones...you can cleanse your own energy field.' This is the sort of thing you will find if you enter the word 'crystals' into Google. Unfortunately this type of pseudoscientific new-age gobbledygook about crystals has become all too pervasive. I recently bought a lovely large crystal of Iceland spar (calcite) for my collection from a shop. The shop assistant advised me to keep it near me when I go to bed, as this would bring harmony and a good night's sleep. Apparently it clears away 'negative energies'. I tried it. It didn't. I often have to tell people that crystals are among the 'deadest' objects in the universe—no auras, no energy fields, no chakras. They don't always like to be told such things.

But this does illustrate a fascination with crystals that goes back a long way into the ancient past. We know that Peking man collected rock crystal quartz, probably to make tools, or possibly for a primitive animistic belief. Australian aborigines also used rock crystal and amethyst as rain stones in rain-making rites, and they also attributed malevolent powers to crystals. The earliest recorded observation of six-sided snow crystals was in China in 135 BC, where they were compared with the pentagonal symmetry of flowers by the author Han Ying in his book *Disconnection* (韓詩外傳).

Many other ancient civilizations recognized crystals as something special, and they were usually used as gem-stones for decoration and for religious purposes. The ancient Egyptians used to carve rock crystal to make venerated objects. The ancient Greeks too were fascinated by crystals, especially the atomists who believed everything was made from particles. Attempts at rational explanations of crystals were made by a number of thinkers at the time, such as Democritus (*c*.470–400 BC), Epicurus (347–270 BC), Lucretius (*c*.95–55 BC), Aristotle (384–322 BC), and the Stoics. Lucretius, for example, explained that the hardness of diamond must be due to branch-like atoms intertwined with one another—not too far in fact from what we know today. Rock crystal was thought for a long time to be made from ice that was so frozen that it would not thaw out. The Roman Pliny the Elder (AD 23–79) wrote 'a violently contracting coldness forms the rock crystal in the same way as ice'.

Most people are unaware that crystals are to be found everywhere: for example, in your bones, teeth, muscle, in building materials, in the earth, in your mobile phone, and in almost every solid object that you can see around you. Even chocolate contains crystals of cocoa butter. Crystals can be colourful, symmetrical, beautiful, and large. For instance, huge gypsum crystals measuring many metres in length were discovered in the Nazca Caves in Mexico as recently as 2000, and a crystal of the mineral beryl was found in the Malagasy Republic that is eighteen metres long and weighs about 380 tonnes.

There is even some evidence that iron crystals many kilometres long may exist at the centre of the Earth. A recent study by Tao Wang, Xiaodong Song, and Han Xia, published in the journal *Nature Geoscience*, reported that the inner core of the Earth, once thought to be a solid ball of iron, has complex structural properties. The team found a distinct inner-inner core, about half the diameter of the whole inner core. The iron crystals in the outer layer of the inner core are aligned directionally, north–south. However, in the inner-inner core, the iron crystals point roughly east–west. Not only are the iron crystals in the inner-inner core aligned differently, but they behave differently from their counterparts in the outer-inner core. This means that the inner-inner core could be made of another type of crystal or phase. But crystals can also be minute, lacking in obvious symmetry and form, and difficult to see without the aid of a microscope.

Crystallography, or the science of crystals, is today of crucial importance in many, often unrecognized, ways. Understanding the nature of crystals, especially their atomic structure, is vital for many practising scientists and for industry. For instance, chemists rely on a knowledge of crystal structure during the process of discovering and synthesizing new chemical compounds in order to identify a substance and then to alter its properties. Similarly, most of the drugs and pharmaceuticals, so vital for health today, at some stage involve crystallographic knowledge as a way to make useful modifications. Pharmaceutical companies usually include in their patents some sort of crystallographic analysis to support their claims; typically, a particular drug is patent-protected by supplying a powder diffraction pattern, or occasionally a full crystal structure determination. In addition, it is crystallography that is involved in helping research into how drugs target proteins, the molecules that are essential for living organisms to function properly.

Then again, materials scientists routinely use crystallography to study new materials having many industrial applications. Crystalline solids have interesting and useful mechanical, electrical, optical, and magnetic properties. The huge semiconductor industry relies on the growth of large crystals of materials such as silicon or germanium. Crystals of lithium niobate are used extensively in the telecom market, e.g. in mobile telephones and optical modulators. It is an excellent material for manufacture of optical waveguides. Your watch contains a crystal of quartz, which enables accurate timekeeping. The production and sensing of ultrasonic waves are achieved by certain crystalline materials, called piezoelectrics, and this has uses in medicine and engineering, and in military applications such as SONAR (SOund Navigation And Ranging) systems, used to detect underwater objects.

The search for superconducting and photo-sensitive materials also depends on a knowledge of crystals and their structures. Similarly, in order to understand how proteins work in living organisms, crystallographic methods are routinely used to determine the shapes and atomic arrangements of the protein molecules and hence understand how they function. Our ability to create fresh antibiotics relies on this knowledge. In the early 1940s, for instance, it was Dorothy Hodgkin's determination of the crystal structure of penicillin that enabled chemists to produce new types of antibiotics. Her work on vitamin B12, and later on insulin, was of enormous importance in understanding these substances and how they work in the body. Even viruses can be made to crystallize, enabling crystallographers to study their molecular structures and hence how they work. One could go on and on listing all the uses of crystallography.

If it were not for crystallography, and especially the discovery of X-ray crystallography in 1912, the world today would look very different. There would not be much of a pharmaceuticals industry, nor useful drugs, no electronics to speak of, no computers, no worldwide communications, no mobile phones, and so on.

I have always argued that crystallography is a scientific discipline in its own right, rather like, say, chemistry or physics, and should not be regarded merely as a technique. Crystallographers have their own international union, their own systems of nomenclature and notation, and their own experimental techniques.

Crystallography is one of the most interdisciplinary subjects in science: crystallographers can be found in chemistry, physics, biology, engineering, materials science, and mathematics departments, as well as in many industries.

But curiously, despite all this importance, the subject of crystallography is relatively unknown, not only by the general public, but even by many scientists. We have all seen science programmes on television on cosmology, particle physics, chemistry, medicine, and so many other scientific topics, but virtually nothing has been shown about crystals or crystallography. Despite having earned something like twenty-six Nobel prizes, crystallography seems to be a 'hidden' subject.

I am most grateful to the following colleagues for reading the text of this book and correcting my many misconceptions: Stephen Blundell (Oxford Physics), Stephen Curry (Imperial College), and Elspeth Garman (Oxford Biochemistry).

Preface

List of illustrations

1 Hexagonal snow crystal and stacking of spheres by Kepler **2**
©Kichigin/Shutterstock.com

2 Images from Hooke's **3**
AA 82 Art. Fig. 1 & 2 (SCH VII). The Bodleian Library, University of Oxford

3 Schematic drawing of the cross-section of two quartz crystals **4**

4 Drawings from Haüy's *Traité de Minéralogie*, and portrait of René-Just Haüy (1784) **6**
From http://reference.iucr.org/dictionary/Law_of_rational_indices. Photo: 8° G 57 BS (Vol. 5), Fig 13 pl. II, Fig 16, Pl. II, Fig 17, Pl. III. The Bodleian Library, University of Oxford

5 Laue, Friedrich, and Knipping's experiment **14**
Based on illustration in Ewald, P.P. (1962), *Fifty Years of X-ray Diffraction*, published for the International Union of Crystallography by N.V.A. Oosthoek's Uitgeversamaatschappij, Utrecht, The Netherlands.

Inset from Friedrich, W., Knipping, P. & Laue, M. In Sitzungsberichte der Math. Phys. Klasse (Kgl.) Bayerische Akademie der Wissenschaften 303–322 (1912)

6 Bragg's Law **16**

7 Left: William Henry Bragg at his ionization spectrometer. Right: a modern diffractometer **18**
©Heritage Images/Glow Images. com; A SuperNova AS2 diffractometer. Image courtesy of Rigaku Oxford Diffraction, a part of Rigaku Corporation, Tokyo, Japan

8 Max von Laue; and William Lawrence Bragg and William Henry Bragg at the British Association Meeting in Toronto (1924) **20**
Les Prix Nobel, 1914. Nobel foundation/Wikimedia Commons/Public Domain; Smithsonian Institution Archives. Image # SIA2007-0340

9 Examples of symmetry in two dimensions **25**

10 Two enantiomorphic crystals of quartz (SiO_2); representation of rotational symmetry; stereogram; effect of symmetry **26**

11 Examples of triclinic crystals **29**

12 Miller indices of a plane **30**

13 Building a crystal structure **33**

14 The fourteen Bravais lattices **35**

15 Packing of spheres **40**

16 Some basic inorganic crystal structures **42**

17 Examples of organic compounds **47**

18 Protein structures **50**

19 Mirror furnace and example of a crystal of $CoSi_2O_4$ **61**
Courtesy of Dharmalingam Prabhakaran, Clarendon Laboratory, Oxford

20 Construction of the reciprocal lattice **65**

21 The Ewald Sphere construction **66**

22 Formation of a Laue pattern by a continuum of wavelengths using the Ewald construction, and example of a Laue photograph **68**

23 Example of two waves **71**

24 Fourier Transformation **73**

25 Formation of a crystal diffraction **76**
Image courtesy of T. R. Welberry, ANU, Canberra, Australia

26 Formation of powder diffraction rings **77**

27 Commensurate and incommensurate modulations **80**
Figure of Charlie Chaplin from the archives of Roy Export Company Establishment

28 Example of an electron diffraction pattern **84**
Courtesy of Conradin Beeli, ETH, Zurich

29 Special lines marked on the fat and thin rhombs of a Penrose tiling **85**

30 Crystal structure of aspirin **91**

31 X-ray diffraction pattern for Form II, and simulation of diffuse scattering **92**
E. J. Chan, T. R. Welberry, A. P. Heerdegen, and D. J. Goossens. 2010. 'Diffuse Scattering Study of Aspirin Forms (I) and (II)', *Acta Cryst.*, B66, pp. 696–707. <http://dx.doi.org/10.1107/S0108768110037055>. Reproduced with permission of the International Union of Crystallography

32 Formation of an image by a thin lens **95**

33 Fourier synthesis of the structure of naphthalene **98**
Computed by the program FOURDEM of T.R. Welberry, ANU, Canberra, Australia

34 An early computer calculation of the electron density of a molecule of naphthalene **99**

S.C. Abrahams, J. Monteath Robertson, and J.G. White. 1949. 'The Crystal and Molecular Structure of Naphthalene. II. Structure Investigation by the Triple Fourier Series Method'. *Acta Cryst.*, 2, p. 238. Reproduced with permission of the International Union of Crystallography

35 The development of a Patterson map **101**

36 The charge-flipping algorithm **105**

37 Modern sealed X-ray tube and typical emission spectra **108**

From L. J. Poppe, V. F. Paskevich, J. C. Hathaway, and D. S. Blackwood. 2002. *A Laboratory Manual for X-Ray Powder Diffraction*

US Geological Survey; courtesy of CSIC. <http://www.xtal.iqfr.csic.es/Cristalografia/>

38 Synchrotron radiation **113**

Courtesy of Diamond Light Source and Albert C. Thompson (ed.), *Center for X-Ray Optics Advanced Light Source X-Ray Data Booklet*, (Lawrence Berkeley National Laboratory, University of California, Berkeley, California 94720, Third edition, September 2009)

39 Use of free-electron laser radiation in protein crystallography **117**

Lawrence Livermore National Laboratory

40 Rietveld refinements **121**

Rietveld refinements using the HRPD diffractometer at ISIS in 1987. Top: Aluminium oxide Al2O3 Lower:: benzene C_6H_6. STFC/Bill David

41 Annular bright and dark field imaging, and ABF image of a crystal of $PbZr_{0.53}Ti_{0.47}O_3$ **124**

From K. Baba-kishi and A.M. Glazer. 2014. 'Local Structure of $Pb(Zr_{0.53}Ti_{0.47})O_3$'. *Appl. Cryst.*, 47, pp. 1688–98. <http://dx.doi.org/10.1107/S1600576714019086>. Reproduced with permission of the International Union of Crystallography

Chapter 1
A long history

The old era

Despite such a wide abundance of crystals, it was not until the 17th century, with the move towards modern scientific rationalism, that the first real successes in understanding their nature since the times of the ancient Greeks were attained. It was the era of the Enlightenment that led to the systematic study of crystals or 'crystallography', the term having been coined by a Swiss physician, Maurice Capeller (1685–1769).

One of the first on the scene was Johannes Kepler (1571–1630), he of planetary fame, who studied snow crystals and their symmetry in his 1611 pamphlet *Strena Seu de Nive Sexangula (A New Year's Gift of Hexagonal Snow)*. This led to his statement 'where there is matter, there is geometry'. He also worked on another related problem, originally said to have been raised by Sir Walter Raleigh and other ships' captains, on the question of the packing of cannonballs on the deck of a ship (presumably to stop them rolling about). Kepler conjectured that the densest, and hence the most stable, arrangements would be in so-called cubic and hexagonal close packings (this apparently simple suggestion remained unproved mathematically until 2003!). You can see similar packings of oranges or apples in a grocery shop. We now know that many atomic arrangements in crystal structures of elements are based on these types of arrangements (Figure 1).

Another dabbler in early crystallography was Robert Hooke (1635–1703) who, in his book *Micrographia*, studied crystals from urine. In Hooke's own words 'Tasting several cleer pieces of this Ice [meaning urine crystals], I could not find any Urinous taste in them, but those few I tasted, seem'd as insipid as water.' He found that he could explain their flat surfaces in terms of the repeated stacking of spheres (Figure 2). He did not know what these spheres actually were, and he was worried about the voids between them; but Hooke was one of the first to consider the concept of periodicity in connection with crystals. He did think that the gaps might be filled by a fluid.

A slight modification was made by Christian Huygens (1629–1695). He was probably aware of Hooke's packing of spherical particles, and suggested ellipsoidal particles instead. René Antoine de Réaumur (1683–1757), who investigated the crystallinity of metal

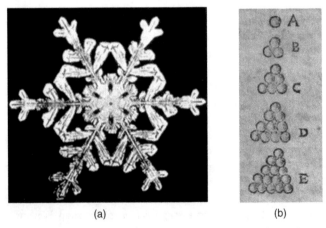

(a)   (b)

1(a). **Hexagonal snow crystal; and (b). stacking of spheres by Kepler.**

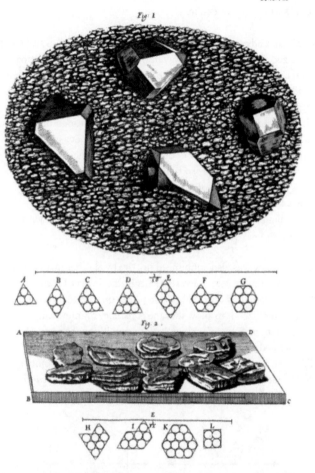

A long history

2. Images from Hooke's *Micrographia* (1665). Top: samples of flint (these are not in fact crystals). Bottom: crystals from urine with packing of spheres to simulate external shapes.

crystals, also believed that in minerals crystallinity arose from some sort of juice or essence.

In another advance, the Danish Bishop Nicolas Steno (1638–1686) theorized in 1669 about the way in which crystals grew, starting from a small seed, although he admitted that he did not know how or why this seed arose. In particular, he explained that mineral crystals grew by addition of particles from an external fluid. From his work, it followed that the angles between corresponding faces on quartz crystals are then identical for all specimens of the same mineral—this was the Law of Constancy of Interfacial Angles, namely *angles between corresponding faces on crystals are the same for all specimens of the same mineral* (see Figure 3). The actual sizes of the faces, on the other hand, are merely consequences of the rates of growth of the different faces, with the largest faces growing most slowly.

A similar law was also announced in 1783 by another investigator of crystal growth, Jean Baptiste Louis Romé de l'Isle (1736–1790). Actually Steno had applied his idea just to quartz, whereas de l'Isle generalized it to show that the constancy of interfacial angles was a characteristic of *any* particular crystalline substance. He also defined a crystal as any body of the mineral kingdom that displayed a polyhedral and geometric shape. His measurements of interfacial angles were enabled by the invention of the *contact*

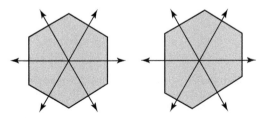

3. **Schematic drawing of the cross-section of two quartz crystals having edges of different size. Directions perpendicular to the sides are 60° from each other, thus highlighting 6-fold symmetry.**

goniometer by his student, Arnould Carangeot (1742–1806). This device measures the interfacial angle using a protractor with one or two rotating arms placed across the interface of two crystal faces of interest. Although simple, and in today's eyes obvious, this instrument and others like it became a vital tool in use until the 20th century for the study of crystal shapes.

But the real breakthrough came in France by the brilliant work of the Abbé René-Just Haüy (1743–1822), born in Saint-Just-en-Chaussée, France, the son of a weaver (Figure 4). Despite his poor background, following his family's move to Paris around 1750 Haüy managed to enter the University of Paris, where he was an outstanding student. His interest in crystallography is said to have arisen by a chance incident. In one story, which parallels the observation of a falling apple by Newton, he was visiting a friend, said to be M. de France du Croisset, who had a collection of crystals. On picking up a prismatic crystal of calcite (chemically calcium carbonate) he accidentally dropped it and noticed that the small broken fragments were recognizably the same shape as other crystals of calcite. He returned to his laboratory in some excitement and began smashing many crystals, observing that no matter how small the fragments, they retained their original form.

According to Georges Cuvier (1769–1832) Haüy cried out, 'Tout est trouvé!' This story may be apocryphal and indeed there are suggestions that Haüy may in fact have built upon the earlier work of Torben Bergman (1735–1784). In 1773 he had published his analysis of the cleavage of calcite in the form of a rhombohedron. There is even evidence that earlier still, at the turn of the 18th century, Domenico Guglielmini (1655–1710) had suggested that cleavage fragments represented primitive polyhedra from which crystals were constructed. Whatever the origin, this led Haüy, who may have been unaware of the earlier suggestions, to postulate that crystals must be made up of regular arrangements of polyhedral units. He could not know what these *molécules*

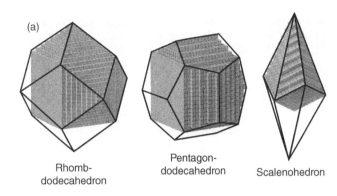

(a)

Rhomb-dodecahedron

Pentagon-dodecahedron

Scalenohedron

(b)

4(a). Drawings from Haüy's *Traité de Minéralogie*, with lines added to emphasize external crystal habits; (b). portrait of René-Just Haüy (1784).

intégrantes, as he termed them, or their constituent parts, were made of at the time.

Today we call periodic polyhedra in a crystal 'unit cells', regions of space that contain groups of atoms that are then repeated to build up the whole crystal. Haüy's observation was probably the first experimental evidence for a type of symmetry that we now call 'translational symmetry': this is the symmetry of repeating patterns, a bit like soldiers on parade. Another, more philosophical consequence of this work was that it demonstrated that solid matter was not homogenous down to the smallest scale, but was instead heterogeneous (i.e. lumpy) and, therefore, left voids in the interstices.

During the early part of the French revolution, after the overthrow of the monarchy, Haüy was imprisoned after refusing to swear an oath of allegiance to the new regime then required of members of the clergy. He narrowly escaped the 1792 September massacres of the clergy at the Saint-Firmin seminary because of the intervention of a former pupil, Étienne Geoffroy Saint-Hilaire (1772–1844), who managed to rescue him and several other professors after a few days. Saint-Hilaire tried various means to secure their release, including disguising himself as an official and entering the seminary buildings. He finally succeeded by bringing a ladder to the walls during the night by which twelve priests escaped. Subsequently, in 1795 Haüy was appointed professor of physics at the École Normale Supérieure.

Haüy was firmly convinced of his own ideas, sometimes to the point of being resistant to the ideas of others. Many arguments occurred between him and other scientists; for instance, de l'Isle was a particular critic. A substantial challenge to Haüy's theories was presented in 1819 by Eilhard Mitscherlich (1794–1863), whose Law of Isomorphism stated that chemically related substances show a close similarity in crystal shape. Haüy insisted, on the other hand, that the forms of *molécules intégrantes* meant that they were a specific characteristic of a substance. He is reported as saying: 'if Mitscherlich's theory is correct, mineralogy would be the most wretched of the sciences'. Nonetheless, his central view of periodicity in crystals formed the basis for further developments in the theories of crystal structure, and effectively heralded the concept of the *space lattice*.

After the restoration, Haüy lost his pension and died in poverty after a fall that fractured his femur. His name is honoured in

France as one of seventy-two inscribed on the Eiffel Tower, and most regard him as the true father of crystallography. A thorough account of Haüy and his work can be found in John Burke's book, *Origins of the Science of Crystals*.

The 19th century saw a huge flowering of new theories of crystals, mainly in France and in Germany (where a more mathematical approach took root). Johann Friedrich Christian Hessel (1796–1872) showed in 1830 that all crystals could be classified within thirty-two so-called crystal classes. Gabriel Delafosse (1796–1878) formally devised the concept of the unit cell in 1840. Moritz Ludwig Frankenheim (1801–1869) and Auguste Bravais (1811–1863) demonstrated that the repetition seen in Haüy's work could be represented mathematically by a point lattice. Furthermore, there was only a finite number of unique lattice types within any spatial dimension. They are known today as the fourteen Bravais lattices in three dimensions. In another development Leonhard Sohncke (1842–1897) combined the point lattices together with the symmetry of rotations (we can loosely think of this as rotational symmetry applied to the groups of atoms within each unit cell). He introduced a new type of symmetry element, the screw axis, where an object is rotated about an axis and then translated through a fraction of a unit cell length. This subsequently gave sixty-five so-called space groups (originally he had sixty-six but subsequently two of them were found to be equivalent).

Even the French scientist Louis Pasteur (1822–1895), best known perhaps for his work on diseases caused by microbes and for finding the cure to rabies, originally began with a study of crystals. In particular, he became interested in the way polarized light (polarized light consists of waves in which the vibration of the electric field occurs in a single plane) was rotated in different senses by solutions made by dissolving crystals known as tartrates. In short, he found that he could, under the microscope, separate his crystals into two piles, each pile consisting of crystals whose shape was the mirror image of crystals in the other pile. After that, when he dissolved the crystals into solution, the optical polarization was rotated in one sense using crystals from one pile and in the opposite sense for those from the other pile. This suggested to him that the actual molecules in the crystals must be either left- or right-handed (today we call this phenomenon *chirality*, an important subject in the pharmaceutical industries). His brilliant analysis was presented in 1848 to much acclaim. It has been suggested that he was lucky to have obtained his crystals in the form in which he could separate them into two piles of opposite handedness. His experiments were carried out in the winter, whereas had he done this in the heat of the summer they would not have grown in this way.

Today this discovery might seem to be of purely academic interest, but at the time there was a lot of research into the biological processes of putrefaction and fermentation, where it was known that the associated fluids also showed optical rotation. Pasteur was also aware that many natural substances exhibited chiral properties, and that even the sense of taste depended on the handedness of certain substances. It is probable that this knowledge led him to suggest that the very essence of life must be tied up with the concept of chirality. He could not have known, of course, that a hundred years later the famous double helix of DNA would be discovered, the helices being of the same chirality for all organisms. This in turn gave scientific support for the theory of evolution, as it suggested that all living organisms are descended from a common ancestor. Moreover, this heralded the modern theories of genetics. So much for the current requirement that scientists seeking research funding should predict the impact of their research.

The space group theory of Sohncke was further elaborated by three individuals working more or less at the same time. In 1891, Arthur Moritz Schoenflies (1853–1928) showed that there were 230 possible space groups in three dimensions. He did this by considering the combination of the thirty-two crystal classes together with the fourteen Bravais lattices and including new symmetries such as the screw rotations of Sohncke. Actually, slightly before him, Evgraf Stepanovich Fedorov (1853–1919), who was working in Russia, had developed the 230 space groups independently. Then again, more or less at the same time, one of the last great amateur scientists, William Barlow (1845–1934) in England, also found the same 230 space groups. Furthermore, he developed ideas about crystal structures in terms of the packing of spheres in different arrangements in order to explain particular crystal forms and chemical compositions. Barlow was apparently quite eccentric and was known to buy up large numbers of white gloves from local shops. According to one story, after his death many of the gloves were found on the walls of his house, arranged in patterns corresponding to the space group symmetries.

Thus, by the end of the 19th century and into the beginning of the 20th, the state of knowledge about crystals was mainly theoretical. Even the concept of atoms at that time was not fully accepted, and how these atoms, if they existed at all, were arranged in a crystal was not known except for some very simple ideas based on the close packing of spheres. Just how molecules could be arranged in a crystal was not known. The time was ripe for a breakthrough to usher in a new era in crystallography.

Dawn of a new era

The breakthrough came in 1895 in Würzburg when Wilhelm Conrad Roentgen (1845–1923) found that a high-voltage electrical discharge tube that he was using seemed to give off mysterious rays that penetrated through materials. He had accidentally discovered X-rays, and this launched an immediate and furious amount of research to find out what they actually were. Roentgen was the first recipient of the Nobel Prize in Physics in 1901. In the early part of the 20th century, there were two apparently conflicting theories, namely that X-rays consisted of beams of neutral particles, or that they were waves. Some experiments, such as those using the famous Wilson cloud chamber, showed tracks corresponding to the passage of particles. Other experiments seemed to suggest waves. It is probably true though, to say that most scientists believed X-rays were waves.

Into this debate strode William Henry Bragg (1862–1942), whom I shall henceforth call WHB. He was convinced that X-rays were beams of neutral particles. Born in Cumbria, he had studied mathematics in Cambridge, graduating with first-class honours. Shortly after, following a recommendation by J. J. Thomson (1856–1940), who in 1906 was awarded the Nobel Prize in Physics for his work on discharge tubes (and incidentally discovered the electron), he took up the then vacant Professorship of Mathematics and Physics at the University of Adelaide, Australia. Knowing very little of physics he had to teach himself, and towards the end of the 19th century began research into alpha particles and into the newly discovered X-rays. In Adelaide he married Gwendoline Todd, daughter of Sir Charles Todd, Government Astronomer and Postmaster General and Superintendent of Telegraphs for South Australia. They had three children, one of whom was named William Lawrence Bragg (1890–1971), henceforth WLB. The young WLB soon showed a precocious scientific ability, putting him far ahead of other boys in his school. The family moved to England in 1907 so that WHB could take up a professorship in physics at the University of Leeds. WLB, despite already having a degree from Adelaide, went to Cambridge to study mathematics, graduating with first-class honours in 1912. While the arguments over particles and waves raged on, a most important event occurred early that year that can truly be said to have changed the world of science.

In Munich the theoretical physicist Arnold Sommerfeld (1868–1951) had established the Institute of Theoretical Physics. Much of the interest of this institute was in the particle-wave problem.

Roentgen had also moved to Munich to continue with his experiments on X-rays. Joining Sommerfeld as *Privatdozent* (an academic title conferred in German-speaking universities) was the young Max Theodor Felix Laue (later to acquire a 'von' in his name when his father was ennobled in 1913). Other personnel at the time included Paul Peter Ewald (1888–1985), a research student under Sommerfeld, and working on the propagation of radiation through solids. At the same time, Paul Karl Moritz Knipping (1883–1935) was a student of Roentgen's and Walter Friedrich (1883–1968) was employed as an assistant in Sommerfeld's institute.

According to the story, one day Ewald happened to mention to Laue that crystals were thought to consist of periodic arrays of atoms. Laue in turn asked him what the typical repeat distances would be in a crystal and was told that they would probably be of the order of Ångstroms (1 Å = 10^{-10} m). Laue himself later said that an idea came to him in a flash after Ewald's comments: if one thought of a crystal as a three-dimensional periodic array, perhaps it could act as a kind of diffraction grating if radiation of the appropriate wavelength were used. Diffraction is the scattering of light when passing through slits or holes or around objects. If the holes are in a regular array (a diffraction grating), the light bends in such a way as to form a periodic pattern on a screen.

Now it was thought, because of some early experiments with X-rays passing through slits, that if they were waves they would have wavelengths in the Ångstrom region. So Laue's brilliant idea was: why not see if crystals could diffract X-rays? However, it seems that when Laue went to Sommerfeld to ask for resources to carry out the experiment, Sommerfeld refused to divert his people towards what he considered to be a waste of time. It is not clear why he thought this. One suggestion made later by Ewald himself was that Sommerfeld would have argued that the thermal vibrations of the atoms would wipe out any possibility of seeing diffraction. Another more recent suggestion is that Laue's description of the diffraction process was patently flawed, since he thought the incident X-rays would excite secondary radiation from certain atoms in the crystal and that it would be this radiation that would be diffracted rather than the incident beam. It has been argued that Sommerfeld would have considered that this would not give rise to diffraction.

In any case, it seems that Laue managed to persuade Friedrich and Knipping to try the experiments in secret. Ewald once told me that to do this experiment, they actually stole Roentgen's X-ray apparatus and carried out the initial experiments at night: after several thwarted attempts, they attained success, probably on 21 April 1912. The first crystal that was used was copper sulfate pentahydrate, a blue crystal familiar to all of us who have grown crystals using a home chemistry set. It was chosen because it was known that copper would give rise to strong secondary radiation (fluorescence) when exposed to X-rays. After a great deal of experimentation, they finally obtained a rough photograph showing the main X-ray beam at the centre but with a few vague blobs away from the main beam—apparent evidence of diffraction (Figure 5). They then used a crystal of zinc sulfide, ZnS, which gave many sharp spots on the film, and moreover, when the crystal was in a particular orientation showed evidence of a 4-fold symmetry. Sommerfeld is said to have thrown a party to celebrate this great discovery, although he did not invite Laue, such was the bad feeling between the two of them at the time.

It was then necessary for Laue to explain the patterns of spots on the film. He derived a set of equations, today called the Laue equations, and with these he could explain many of the spots on the ZnS pattern, but not all. It appears that he made some incorrect assumptions.

First of all, he persisted, for reasons that cannot be understood, with his idea of secondary radiation. He should have realized that this was incorrect, because Friedrich and Knipping had also

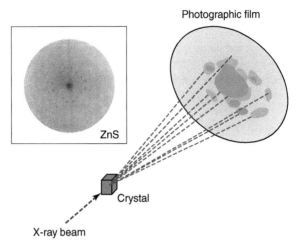

5. Laue, Friedrich, and Knipping's experiment in April 1912 with a copy of the original pattern obtained from a crystal of copper sulfate pentahydrate. Inset shows the Laue pattern from zinc sulfide crystal.

obtained a good pattern from a crystal of diamond. Diamond is made of carbon atoms, and such light atoms would not be expected to exhibit significant fluorescence. The result was that he tried to explain the pattern assuming that the X-rays were of a single wavelength; later, in order to fit the pattern better, he used up to five distinct wavelengths, but even this did not work. Moreover, Laue's knowledge of the various theories of crystal structure in force at the time was weak, as he later admitted. He assumed that in ZnS the molecules would lie at the corners of cubic unit cells: this is what is sometimes called a simple cubic structure—actually, this is a highly unlikely arrangement of atoms: only crystals of polonium crystallize in such an arrangement under normal conditions.

At this point, Lars Vegard (1880–1963), a Norwegian scientist who had studied with WHB, happened to be in Munich and heard about Laue's discovery. He wrote a now-famous letter to WHB with a detailed description of the German experiment, and naturally WHB became deeply interested in it. Being wedded at the time to his particle theory of X-rays, he felt that he could explain Laue's patterns, not so much as waves but as particles travelling down 'avenues' between atoms in the crystal. By this time, WLB had joined him briefly in Leeds and together they set about trying some experiments to prove the point. The experiments failed.

Returning to Cambridge, WLB continued to ponder this problem and then suddenly, while walking behind King's College, realized that he could explain Laue's photographs after all. The German experiments did indeed show X-rays were waves. WLB had had a good grounding in the theory of optics from lectures by C. T. R. Wilson (1869–1959) and from reading the 1909 book *An Introduction to the Theory of Optics* by Arthur Schuster (1851–1934).

He noticed in particular that the spots on the film were elliptical in shape and became flatter when the film was moved further away from the crystal, suggesting some kind of reflection of the X-ray beam. He knew that the theories of crystal structures meant that the atoms should lie on parallel planes, visible in any direction. He supposed then that an incident X-ray beam could be thought of as being reflected by a particular set of planes, acting rather like mirrors, but with the difference that the reflected beams could interfere with one another to give either constructive or destructive interference, depending on whether the crests of the waves were in phase or out of phase. This then suggested a simple equation to calculate the positions of the spots (we now call them *reflections*) on the film. Originally, it was in the form

$$n\lambda = 2d\cos\theta$$

where λ is the wavelength, d the spacing between the planes, and θ the angle the X-ray beam makes with the perpendicular to the planes (n is an integer). It was soon changed to

$$n\lambda = 2d\sin\theta$$

by redefining θ to be the angle the X-ray beam makes with the planes, and is known as Bragg's Law in the form used today (Figure 6). A reflection occurs only if all three quantities satisfy this equation simultaneously; otherwise, there is no diffraction spot on the film. WLB also realized that the X-rays used by Friedrich and Knipping must have been white, i.e. they consisted of a continuum of wavelengths rather than of a few specific wavelengths.

Laue had considered only a few wavelengths were present in the X-ray beam because he thought a white beam would cause uniform fogging of the film. But Bragg's Law shows that only those values of the wavelength λ that satisfy it simultaneously with respect to d and θ are automatically picked out.

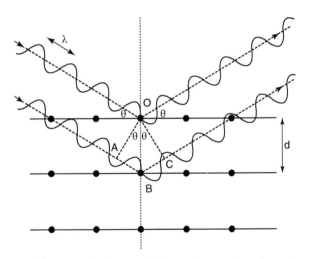

6. **Bragg's Law. ABC is the phase difference between incoming and outgoing waves. For constructive interference this must equal a whole number of wavelengths.**

In addition to this, WLB also found that he could explain the complete ZnS diffraction pattern by assuming that the molecules were not only on the corners of a cubic unit cell but also occupied the centres of each face, a so-called face-centred cubic array. In this, he was fortunate in having been introduced to the very simple models of crystal structures then being advanced by William Pope (1870–1939), Professor of Chemistry at Cambridge, and by William Barlow.

WLB's paper describing his theory was read to the Cambridge Philosophical Society by J. J. Thomson on 11 November 1912, and published soon after. It can truly be said that this first paper began the modern field of X-ray crystallography—not bad for a mere 22-year-old!

But it did not stop there. WHB was now convinced by his son's theory that X-rays were indeed waves. Actually, he was ahead of his time in suggesting that they were, in fact, both waves and particles depending on what sort of experiment was done. This was an idea that was accepted in 1924 with the wave–particle duality hypothesis of Louis-Victor-Pierre-Raymond, 7th duc de Broglie (1892–1987). WHB used to say: 'On Mondays, Wednesdays, and Fridays we use the wave theory; on Tuesdays, Thursdays, and Saturdays we think in streams of flying energy quanta or corpuscles.'

Throughout 1913 and into 1914 WHB then began working with his son to use the new method to study the structures of many crystals, starting with the simplest and going on to ever more complex cases. WLB remembered in his Nobel lecture: 'It was a wonderful time, like discovering a new goldfield where nuggets could be picked up on the ground, with thrilling new results every week.' A most important development during this time was WHB's introduction of the ionization spectrometer, constructed by his instrument-maker C. H. Jenkinson in Leeds. This device enabled the intensity of each X-ray reflection to be measured precisely by scanning through each intensity maximum, and was a considerable advance over the Laue method. I suggest that it was the use of this spectrometer that led to the change in Bragg's formula from the cosine to the sine form as it is natural with a spectrometer to measure all angles from the incident beam direction out towards higher angles. The spectrometer was the forerunner of the modern diffractometer used today in crystallography laboratories all over the world (Figure 7).

In 1913, WLB used the spectrometer and the Laue method to derive the crystal structures of the alkali halides. This was the first complete determination of a crystal structure. His model for the crystal structure of sodium chloride, common salt, had the sodium and chlorine atoms equally spaced and alternating in all directions rather like the squares on a chessboard. It is fascinating to realize that this was a highly controversial model of the structure: many chemists did not like it as they had expected that the two ions would form molecules. At this time, the concept of ionic bonding was not developed. For example, the Professor of Chemistry at Leeds, Arthur Smithells (1860–1939), even begged WLB to 'please make the sodium a little closer to the chlorine'!

7. Left: William Henry Bragg at his ionization spectrometer. Right: a modern diffractometer.

The sense of opposition to WLB's model of the structure can also be gauged by the following amusing letter to *Nature* by the English chemist H. E. Armstrong as late as 1927:

Poor Common Salt!
'Some books are lies frae end to end' says Burns. Scientific (save the mark) speculation would seem to be on the way to this state! ...Professor W.L. Bragg asserts that 'In Sodium Chloride there appear to be no molecules represented by NaCl. The equality in the number of sodium and chlorine atoms is arrived at by a chess-board pattern of these atoms; it is a result of geometry and not of a pairing of the atoms'.

This statement is more than 'repugnant to common sense.' It is absurd to the n...[th] degree, not chemical cricket. Chemistry is neither chess nor geometry, whatever X-ray physics might be. Such unjustified aspersion of the molecular character of our most necessary condiment must not be allowed any longer to pass unchallenged. A little study of the Apostle Paul may be recommended to Prof. Bragg, as a necessary preliminary even to X-ray work, especially as the doctrine has been insistently advocated at the recent Flat Races at Leeds, that science is the pursuit of truth. It were time that chemists took charge of chemistry once more and protected neophytes against the worship of false gods; at least taught them to ask for something more than chess-board evidence.

It is interesting to note that William Barlow in 1883 had suggested that alkali halides would adopt WLB's model with its alternating atoms. And, curiously, the School of Chemistry Collection at the University of Edinburgh has a model made by Alexander Crum Brown (1838–1922), in apparently the same year, in which he used balls of wool of alternating colour resembling the structure of sodium chloride. It is not known if they were aware of each other's ideas.

Following the determination of the structure of salt, WHB and WLB published the crystal structure of diamond, a most important structure, especially as it has essentially the same arrangement of atoms as in silicon and germanium. The development of modern electronics has depended on this knowledge.

Father and son continued to work feverishly together to develop the subject more fully, deriving the structures of ever more complex crystals. But the work came to a halt with the beginning of World War I in 1914. WHB went to work for the Admiralty on the use of hydrophones to detect submarines, while WLB went to the front in France to join a special unit involved in sound ranging as a method of detecting the positions of enemy guns. This was particularly successful, and the work that was done by this unit under WLB is credited with having made a significant contribution to shortening the war in 1918.

During the war, Laue, by now von Laue, was awarded the 1914 Nobel Prize in Physics, with the 1915 prize going to WHB and WLB. The Braggs are the only father and son team to share a Nobel Prize, and the young WLB, at the age of twenty-five, remains the youngest ever Nobel winner for science (Figure 8).

8. Left: Max von Laue. Right: William Lawrence Bragg and William Henry Bragg at the British Association Meeting in Toronto (1924).

After the war, the Braggs returned to their research into crystals. WHB established a research group concentrating mainly on crystals containing organic molecules, while WLB, in order not to cut across his father's work, went to Manchester to work on metals and inorganic materials. Interestingly, the Braggs encouraged women in science, something that was rare in those days. In fact, out of eighteen of WHB's students, eleven were female. Among these was Kathleen Lonsdale (1903–1971), who became one of the two first female Fellows of the Royal Society. She solved the structure of hexamethylbenzene, showing that the hexagonal ring of carbon atoms was flat—an important finding for the chemistry of aromatic compounds. The formidable John Desmond Bernal (1901–1971) joined WHB's group in 1922, and he also encouraged female scientists, most notably his student Dorothy (Crowfoot) Hodgkin (1910–1994), who was to be awarded the Nobel Prize for the structure determinations of penicillin and vitamin B12 in 1964. As if this was not enough, Dorothy went on to head a team that solved the structure of insulin in 1969. Another well-known student of Bernal was Helen Megaw (1907–2002) who worked on the structures of ice, minerals such as feldspar, and on materials with important electrical properties. Megaw Island in Antarctica was named after her in honour of her work on ice.

And so the field of crystallography continued to flourish, not only in Britain but in many countries throughout the world, with many important discoveries. WLB moved to the Cavendish Laboratory in Cambridge in 1938, and there established a highly successful crystallography laboratory. He worked together with Max Perutz (1914–2002) who, together with John Kendrew (1917–1997), solved the crystal structures of myoglobin and haemoglobin (the molecules in blood that transport oxygen). These were the first protein structures to be solved and gained Perutz and Kendrew the Nobel Prize in 1962. More or less at the same time in WLB's laboratory, James Watson (1928–) and Francis Crick (1916–2004) worked out the structure of the molecule of DNA, ushering in the modern science of genetics. This work resulted from collaboration with Maurice Wilkins (1916–2004) and Rosalind Franklin (1920–1958), with the then PhD student Raymond Gosling (1926–2015) at King's College London. It was Franklin's famous X-ray photograph 51 that gave Crick and Watson the proof that DNA consisted of a double helix. Crick, Watson, and Wilkins shared the Nobel Prize in 1962, Franklin having died by this time, thus making her ineligible for the prize.

William (Bill) Astbury (1898–1961), who originally worked with WHB and then went to Leeds, and was a pioneer of the science of molecular biology, actually obtained a similar photograph to that of Franklin one year earlier. However, he failed to see its significance, possibly because he had been suffering a period of disillusionment at the time after being turned down for a major research grant.

After retiring from the Cavendish Laboratory, WLB took up the directorship of the Royal Institution of Great Britain, where he built yet another important research group. Here, the second protein structure and the first enzyme structure, that of lysozyme, was worked out by David Chilton Phillips (1924–1999) and others in 1965. Lysozyme is found in egg white and in tears, and is responsible for destroying harmful bacteria, and so is an important part of the immune system.

Important advances in crystallography have also occurred in many other countries. For instance, in the US, Linus Carl Pauling (1901–1994) made important contributions to the understanding of the way atoms bond together and showed that protein structures contain helical arrangements of amino acids. Pauling obtained the Nobel Prize in Chemistry in 1954 (he also obtained the Nobel Peace Prize in 1962).

In more recent times crystallography has continued to advance both in the invention of new technologies (radiation sources, detectors, environmental conditions, etc.) as well as in the ability to solve ever more complex structures. Nowadays, the structures of biologically important substances containing thousands of atoms are routinely solved, something that could not have been foreseen by the early workers in structure determination. Crystallography laboratories can be found throughout the world, in universities, industries, and in research institutes—everywhere that knowledge about the atomic structures of solids is needed. According to the International Union of Crystallography (IUCr), there are at present approximately 30,000 crystallographers around the world.

Chapter 2
Symmetry

Concepts

In order to explain what crystals are and how their structures are described, we need to understand the role of *symmetry*, for this lies at the heart of crystallography. We all have some innate idea of what we mean by symmetry, but formally it is the property of an object whereby it appears to be unchanged after some sort of transformation has been applied to it. Thus, if a rotation about an axis leaves the object looking like nothing has happened (see Figure 9), the object is said to exhibit rotational symmetry. Even the word SWIMS looks the same upside down as it has a 2-fold rotational symmetry, that is, it can be rotated through 180º and still looks the same. Similarly, if the object can be reflected across a plane, leaving it unchanged, we talk about mirror or reflection symmetry. Symmetry often appears in combinations too: thus the second and fifth images from the left at the top in Figure 9 exhibit both rotational and mirror symmetry. Note that the fourth image does not possess mirror symmetry, but it does have a 4-fold rotation axis.

These examples are of what we call *point symmetry*. In other words, they are described by symmetry operations acting through a point. But what has this got to do with crystals? Crystals too exhibit point symmetry, and it is found that in

2-fold   2-fold   3-fold   4-fold   6-fold

9. Examples of symmetry in two dimensions. Top: rotations about an axis normal to the page. Bottom: reflections across a vertical plane.

classical crystals there are thirty-two possible so-called *classes* in three dimensions and seventeen in two dimensions. These consist of combinations of symmetry operations to form what in mathematics are called *groups*. For instance, consider the symmetry of quartz crystals (Figure 10(a)): quartz is a great favourite in the crystal healing community, which seems to believe that it has mysterious properties. This crystal has several flat faces, labelled here *m*, *r*, and *z*. As an aside, notice that the crystal on the right is the mirror image of the one on the left: they are said to be *enantiomorphically* related by a reflection plane and are not made equivalent by any rotation (the same is true for your left and right hands—try to turn your left hand into the right hand by simple rotation). This type of what is called *chiral symmetry* gives quartz an interesting property. Some quartz crystals can rotate the plane of polarization of light one way and others the opposite way, rather like Pasteur's tartrates, and, just as he found, we see that the actual crystals are mirror

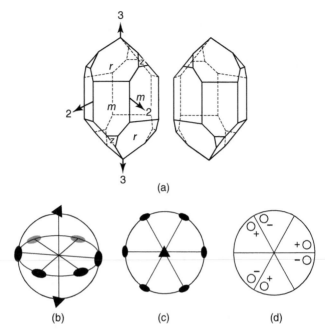

10(a). **Two enantiomorphic crystals of quartz (SiO₂); (b). representation of rotational symmetry; (c). stereogram; (d). effect of symmetry.**

images of one another. The same is found when the internal atomic arrangements in the two crystals are examined.

The arrows on the left crystal represent axes of rotation. Thus the axis marked 3 represents a 3-fold rotation, where an object appears unchanged after rotating through successive angles of 120° about an axis, and the number 2 denotes 2-fold rotations. You can see, for example, how the 2-fold rotations relate the faces marked *m*, *r*, and *z*. In Figure 10(b) I have 'thrown the crystal away' and replaced it by a sphere in the same view on which the rotation axes are shown, this time using the symbols adopted by the IUCr. The triangles represent 3-fold axes, and the ellipses 2-fold axes.

Figure 10(c) shows this now looking down the 3-fold axis: this type
of plot is called a *stereographic projection*, or *stereogram* for short.
Notice that there are in total three 2-fold axes related by 120° as a
result of the 3-fold axis: this shows that each symmetry operation
affects all other symmetry operations, one of the properties of a
mathematical group. Figure 10(d) shows what happens to any
object placed in this symmetry group. The circles represent such
an object, for instance, in our example of quartz they could
represent the *m* faces. Alternatively, they could represent atoms,
groups of atoms, or molecules making up the atomic structure of
the crystal. The plus and minus signs indicate whether this object
lies above or below the plane of projection of the diagram. Since
the 2-fold axes lie in this plane any object above is rotated about
the axis to appear below the plane. It can be seen that the
combination of 3- and 2-fold axes has six related *m* faces,
alternatingly pointing up and down. The full list of symmetry
operations for this crystal is:

$$1 \quad 3 \quad 3^2 \quad 2 \quad 2 \quad 2$$

The symbol 1 is the mathematically trivial *identity* operation
which does nothing to the representational object, and is there
merely for completion. The symbol 3^2 means two rotations of
120°, i.e. through a total angle of 240°. Note that $3^3 \equiv 1$, in other
words, it brings the object back to its original position: such
behaviour is a property of a mathematical group. This particular
group of operations is one of the thirty-two possible point groups,
and crystallographers give this group the symbol 32 in the
so-called International Notation. Table 1 lists all thirty-two point
groups in two commonly used notations: Schoenflies and
International.

In addition to classification by point groups, crystals are divided
into seven *crystal systems*, according to which symmetries are
present. Thus, the triclinic crystal system contains two point groups,
1 and $\bar{1}$: the first contains no symmetry apart from identity.

Table 1

Crystal system	Schoen-flies	Interna-tional	Axes restrictions
Triclinic	C_1 $S_2(C_i)$	1 $\overline{1}$	–
Monoclinic	C_2 $C_{1h}(C_S)$ C_{2h}	2 m 2/m	$\alpha = \beta = 90°$
Orthorhombic	$D_2(V)$ C_{2v} $D_{2h}(V_h)$	222 mm2 mmm	$\alpha = \beta = \gamma = 90°$
Tetragonal	C_4 S_4 C_{4h} D_4 C_{4v} $D_{2d}(V_d)$ D_{4h}	4 $\overline{4}$ 4/m 422 4mm $\overline{4}$ 2m 4/mmm	$a = b; \alpha = \beta = \gamma = 90°$
Trigonal	C_3 $S_6(C_{3i})$ D_3 C_{3v} D_{3d}	3 $\overline{3}$ 32 3m $\overline{3}$ m	$a = b; \alpha = \beta = 90° \; \gamma = 120°$
Hexagonal	C_6 C_{3h} C_{6h} D_6 C_{6v} D_{3h} D_{6h}	6 $\overline{6}$ 6/m 622 6mm $\overline{6}$ m2 6/mmm	$a = b; \alpha = \beta = 90° \; \gamma = 120°$
Cubic	T T_h O T_d O_h	23 m $\overline{3}$ 432 $\overline{4}$ 3m m $\overline{3}$ m	$a = b = c; \alpha = \beta = \gamma = 90°$

The second has a line above it; this is the symbol for a *centre of inversion* (also known as a *centre of symmetry*). This is the type of symmetry where any point in the crystal whose coordinates are x,y,z is equivalent to one at –x,–y,–z (Figure 11).

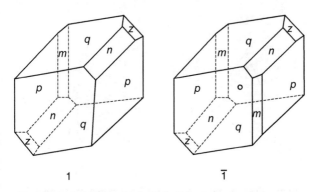

11. Examples of triclinic crystals. Left: point group 1, non-centrosymmetric. Right: point group $\bar{1}$, centrosymmetric.

Notice how one of the small faces marked m is missing from the front of the crystal on the left, thus removing the centre of symmetry.

Suppose now that we try to attach axes of reference to define the faces of these two crystals. Crystallographers label them as a, b, and c (Figure 12) and the angle between these axes as α (between b and c), β (between a and c), and γ (between a and b). It follows as a consequence of this type of triclinic symmetry that there can be no relationships between the axes or the interaxial angles. It is important to understand that any restrictions in the relationships of the axes and angles arise as a consequence of the symmetry present in the crystal and not the other way around. The final column in Table 1 gives the restrictions imposed by the symmetry in each crystal system. In the cubic system, there are always four 3-fold axes of symmetry present, and it is this combination of rotations that interchanges the axes a, b, and c, making them all equal and at right angles to one another.

It is perfectly possible, for instance, to find that measurements made on a crystal seem to indicate, say, a = b with interaxial angles

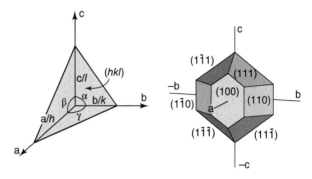

12. Miller indices of a plane. Example of indexing faces of a crystal of adamite (Zn_2AsO_4OH).

all equal to 90°, suggesting tetragonal symmetry. But then, through closer examination by more careful measurements or by looking at the crystal structure (the arrangement of atoms in the structure), it may be seen that there are no 4-fold axes present, as required for the crystal to be tetragonal. The apparent axial and angular equalities are just an accident and not perfectly true. An example that I have worked on myself is the compound lead zirconate, which looks tetragonal from measurements of the axes and angles, but is in fact orthorhombic as found on looking at the arrangement of atoms in the crystal. It is therefore bad practice to define the crystal systems in terms of their axial relationships rather than their symmetries. Many textbooks are misleading on this point.

Miller indices

Now that we see that macroscopic crystals tend to have plane faces connected by symmetry, and that the internal arrangements of molecules lie on planes also related by symmetry, it is important to have some notation to specify these planes. It was William Hallowes Miller (1801–1880), a Welsh mineralogist, who came up with a notation that we call the Miller index.

Consider a plane intersecting the three axes a, b, and c (see Figure 12). Suppose that this plane makes intercepts on these three axes at a/h, b/k, and c/l. It turns out, as a consequence of the ideas of the early crystallographers, that in normal crystals h, k, and l are integers, and according to convention, the plane is then denoted by the Miller index (hkl). Miller indices are used whether one is talking about the actual faces of a macroscopic crystal or of a set of planes of atoms and molecules making up the crystal structure. Following from the original work of WLB and WHB, X-ray reflections, because they arise from scattering by planes, are also labelled hkl, but without parentheses.

Also shown in Figure 12 is the indexing of the faces of the cubic mineral adamite. As an example, consider the large face at the front. This is perpendicular to the a-axis, and is parallel to b and to c. Therefore, it cuts off intercepts on the axes at $(a/h \infty \infty)$, which is equivalent to $(a/h \ b/0 \ c/0)$. This gives the Miller index as $(h00)$ and by convention we take h to be the smallest integer possible to give (100) when indexing actual crystal faces. The (111) face cuts off unit intercepts on all three axes. When labelling X-ray reflections, h, k, and l can take larger values.

Lattices

The use of point symmetry to describe crystals is fine for dealing with the external morphology (shape) of crystals, but when it comes to the internal arrangements of the atoms, i.e. the *crystal structure*, another type of symmetry has to be added. This is called *translational symmetry*, and represents periodicity in space. In fact, this symmetry is the most fundamental type of symmetry found in all normal crystals (we assume here that the crystals are perfect, although it should be understood that, usually, real crystals contain a number of defects and dislocations, which locally break the translational symmetry).

This has important implications for many of the properties (thermal, electrical, mechanical, optical, etc.) of solid materials. Translational symmetry can be found not only in crystals but in many other areas. For instance, bricks making up a wall are stacked together in a regular array which involves repetition in two dimensions. Wallpaper and carpet designs also show translational symmetry.

There is a neat way in which mathematicians represent translational symmetry called a *lattice* (see middle of Figure 13). This is simply a regular repeating array of points. It is a mathematical construct only and has no physical existence. It serves just to act as a kind of template, which tells us how to place atoms and molecules. Therefore, the points in the diagram should not be confused with atoms. Also shown in the diagram is a region, marked A, bounded by full lines. This is a *unit cell*, which has the property of occupying a region of space such that when repeated it fills all space, rather like the way a tile can be repeated to cover a floor, leaving no gaps. There is an infinite number of ways of defining this unit cell: as an example, two other possible choices are shown. B has the same volume as A. Each of them contains one lattice point.

Unit cells that contain a single lattice point are known as *primitive* unit cells, and the lattice can then be called a *primitive* lattice. The unit cell marked C, on the other hand, has twice the volume of a primitive cell: it contains two lattice points (a useful tip to see this is to move the origin of the unit cell slightly and then count the number of lattice points inside, as indicated by the dashed lines). This type of unit cell is called a *centred* cell. Crystallographers use a number of conventions in choosing a particular unit cell description, but in principle, any choice will do.

An important aspect of the symmetry of lattices is that the number of types of rotational symmetry that are possible is limited to 1-, 2-, 3-, 4-, and 6-fold. In conventional lattice theory,

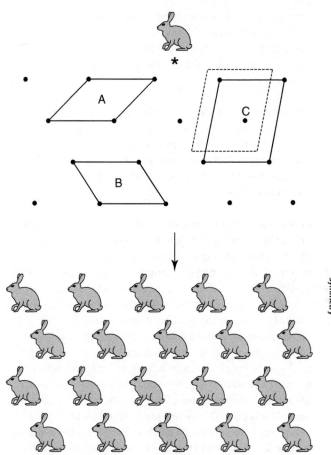

13. Building a crystal structure (bottom) with a molecule (top) and a lattice (middle).

it is impossible to have, say, 5- or 7-fold symmetry. This can be seen if you try to tile a floor with pentagons or heptagons: you will find that this cannot be done without leaving gaps, thus breaking the periodicity needed for a lattice. If you feel wealthy enough try it yourself with a number of polygonal shaped coins (such as the fifty-pence coins used in the UK) which have seven sides. If these coins were hexagonal in shape, then they could be placed on a table, leaving no gaps, and this would make it difficult for the visually impaired to pick them up.

It was shown by Bravais in the 19th century that, when one takes into account the symmetries of the seven crystal systems, there are, in fact, only fourteen unique types of lattice (Figure 14), and any other type of lattice description is equivalent to one of the fourteen by some transformation, such as through a rotation and a redefinition of the axes.

Consider a couple of examples. Remember that the cubic system is defined by the presence of four 3-fold axes of symmetry. In Figure 14 three types of cubic unit cell are shown. The first one is primitive, labelled cP and the four body diagonals coincide with the four 3-fold axes of symmetry. The second has an extra lattice point at the centre of the unit cell, and is known as body-centred, labelled cI. Because the extra point lies on the intersection of the body diagonals, the four 3-fold axes are retained. Similarly, if lattice points are placed at the centres of the cube faces, called all-face-centring cF, this again retains the four 3-fold axes of symmetry. However, suppose we place lattice points at the centres of the top and bottom faces only of the cube, but not on the other faces. This is called C-face centring, provided that we define the c-axis to be perpendicular to this face. This would break the 3-fold rotational symmetries of the cube because there would be no permutation of the cube faces, and so such a type of centring does not belong in the cubic system (even though experimentally in a crystal, we might find that accidentally a = b = c, $\alpha = \beta = \gamma = 90°$).

In the tetragonal system, we see that there are two possible choices, a primitive cell tP and a body-centred cell tI. Both of these preserve the 4-fold axis of symmetry along c defining the tetragonal system. If one adds points at the centres of the top and

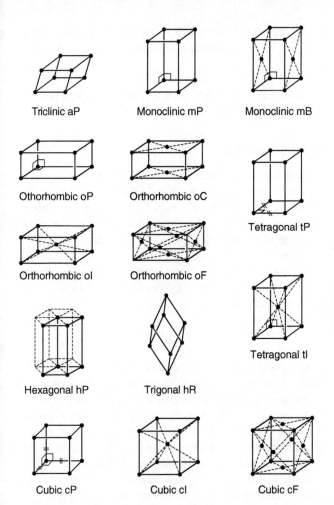

Triclinic aP

Monoclinic mP

Monoclinic mB

Othorhombic oP

Orthorhombic oC

Tetragonal tP

Orthorhombic oI

Orthorhombic oF

Hexagonal hP

Trigonal hR

Tetragonal tI

Cubic cP

Cubic cI

Cubic cF

14. The fourteen Bravais lattices.

bottom faces, this would make a tC-centred cell, which still
preserves the 4-fold axis along c. However, by redefining the a-
and b-axes by rotating through 45° around the c-axis, one obtains
the tP unit cell; and tC, therefore, is not uniquely different, even

though it is perfectly acceptable to use this unit cell description. Similarly, tF is equivalent to tI by the same rotation around the c-axis.

Proceeding in this way, it is found that there are indeed only fourteen unique possibilities.

Crystal structure

Just as a lattice is a regular array of points, a crystal structure consists of a regular array of physical entities: atoms and/or molecules. In Figure 13 (top) we start with a molecule, which I have represented by a rabbit, and then this must be combined in some way with the lattice to generate mathematically the crystal structure (bottom of Figure 13). There is a nice mathematical tool for doing this, called *convolution*. Put simply, the convolution of the molecule M with a lattice L defines a crystal C. I shall write this symbolically as

$$C = M * L.$$

This mathematical operation can be thought of as follows: slide the molecule M over the lattice L and each time the molecule runs over a lattice point fix it there. In this way, a repeating set of molecules is created in sympathy with the lattice. This equation, therefore, can be thought of as a mathematical definition of a crystal structure. Mathematically, the lattice is of infinite extent, and so this would define an infinite crystal. If we want, for example, to define a finite crystal, then we can introduce a mathematical 'shape' function S, which has the property of equalling 1 inside a region and 0 outside. Then

$$C = M * (L \times S).$$

The mathematical operation of multiplication by the shape function ensures that there are no molecules outside S.

Notice that a unit cell can be drawn on the crystal structure, just like on the lattice. In fact, in describing crystal structures, crystallographers will start with a unit cell and then place the atoms within it. The crystal is then created by stacking together copies of the unit cell in three dimensions. You can see from this that in order to describe all the 10^{30} or so atomic positions within a typical crystal it is only necessary to list them within one unit cell and then let the translational symmetry generate all the others. The use of symmetry saves a lot of work.

Space groups

In fact, a more succinct way of generating a description of a crystal structure is used by crystallographers. Suppose that in addition to lattice repeats, the point symmetry of the molecule is taken into account. This means that it is not necessary to specify all the atomic positions in the molecule, but again use symmetry to generate all the atoms in the molecule and then use the lattice to generate the crystal structure. This is how computer programs that draw crystal structures do it. Even more work saved!

This combination of the thirty-two point symmetries and the fourteen Bravais lattices results in the so-called *space groups*. There are seventy-three different types, known as the *symmorphic space groups*. However, it was realized in the 19th century that the combination of point symmetry and lattices gives rise to some new types of symmetry operations, known as screws and glides. Screws involve a rotation plus a translation of an atom through a fraction of a unit cell repeat, to form a helical arrangement. Glides involve a reflection through a mirror followed by a fractional translation. These extra symmetry operations give rise to a further 157 *non-symmorphic* space groups, making a total of 230 altogether.

Therefore, it is found that any normal crystal belongs to one of the 230 space group types. There are two notations in use to denote space groups: one is the so-called International Notation, and the other is due to Schoenflies. Crystallographers almost always use the International Notation. These are all listed in the 'Crystallographer's Bible' known as the International Tables for Crystallography. This has been through a number of revisions under the auspices of the International Tables Commission of the IUCr, the latest versions being known as Volume A. This is a most important work and anyone practising crystallography must become very familiar with its contents.

Chapter 3
Crystal structures

Close packing

Before the breakthroughs of Laue and the Braggs in 1912, our knowledge of crystal structures was formed mainly through argument and conjecture, based on the concepts of symmetry as it was known then and on the external shapes of crystals. Considerations of the packing of spheres proved to be a useful way to describe many simple inorganic structures, especially those of the elements, and were subsequently demonstrated by X-ray diffraction.

Figure 15 shows the way in which spheres (representing atoms of course) can be packed together to form different structural arrangements. Start with a set of spheres in close contact to form a layer marked A. This arrangement has 6-fold symmetry. Now, you should see that there are small gaps between the spheres and that there are, in fact, two types, marked 1 and 2 in the diagram. We now add another layer, B, such that the spheres lie directly on the spaces marked 1. Suppose then that the next layer is exactly the same as layer A and thus we continue to form a repeating structure of layers ABABABAB...etc. Such a structure forms the so-called *hexagonal close-packed structure* (*hcp*), a structure that is found in many elements, such as beryllium, magnesium, and zinc.

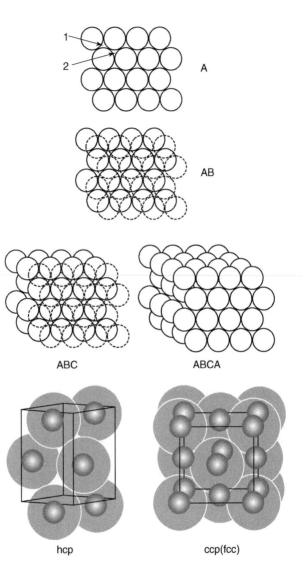

15. **Packing of spheres.**

If, on the other hand, the third layer positions the spheres over the gaps marked 2 to form three layers ABC and then continues as ABCABCABC ... etc., the *cubic close-packed structure* (*ccp*) is obtained, a common arrangement for elements such as copper, silver, and gold. In this structure, the atoms are arranged at the corners of the unit cell and at the face-centres, making a *face-centred cubic structure* (*fcc*), and this is the model arrangement suggested by Kepler for the packing of spheres.

It is important to recall from the work of Kepler that the reason for atoms to pack so closely together is that, in so doing, they form the densest array possible. It seems that atoms (and molecules) want to get as close as they can in order to provide a stable structure. For instance, in *fcc* crystals the triangular faces with 3-fold symmetry tend to be larger than those with 4-fold symmetry: the density of packing of spheres on the triangular faces is approximately 15 per cent higher than on the square faces. This results in the crystals tending naturally to grow in the form of octahedra.

Suppose we place atoms at the corners of a cubic unit cell and an additional atom of the *same type* at the centre of the unit cell. This forms a *body-centred cubic structure* (*bcc*), with a body-centred lattice. This is not a close-packed structure, but nonetheless it is quite common. For instance, it is the structure of molybdenum. On the other hand, suppose the atom at the centre of the unit cell is *not the same* as at the corners (Figure 16(a)). Such a structure, as exemplified by caesium chloride CsCl, is described by a primitive lattice and is not therefore body-centred. Unfortunately some textbooks get terribly mixed up about this and describe it, incorrectly, as having a body-centred lattice.

Figure 16(b) shows another cubic structure, that of common salt. This was the first crystal structure to be determined back in 1913 by WLB, and was at the time very controversial. To generate this structure, consider placing, say, a sodium at the origin of a cubic

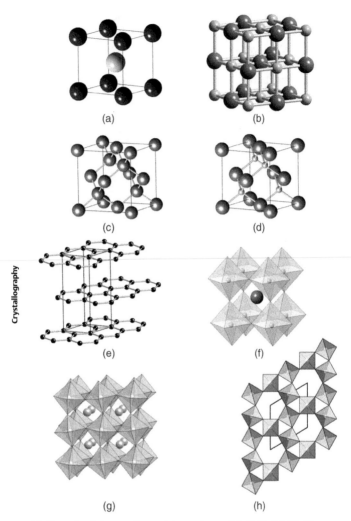

16. Some basic inorganic crystal structures: (a). CsCl; (b). NaCl; (c). diamond; (d). ZnS; (e). graphite; (f). perovskite; (g). SrTiO$_3$; (h) high-temperature quartz.

unit cell, in other words, at a coordinate of (0, 0, 0) with respect to the three axes of the unit cell. Then place a chlorine halfway along the cubic unit cell edge, i.e. at (½, 0, 0). The lattice in the salt structure is the *fcc* lattice which takes each of the atoms and adds fractional coordinates of (0, 0, 0), (½, ½, 0), (½, 0, ½), and (0, ½, ½). This then gives the following positions of the eight atoms in the sodium chloride structure:

$$\text{Na: } (0,0,0),(½,½,0),(½,0,½),(0,½,½)$$
$$\text{Cl: } (½,0,0),(0,½,0),(0,0,½),(½,½,½)$$

For the record, a crystallographer today would describe this structure by using the space group symbol Fm$\overline{3}$m (an example of the International Notation) with Na at position (0, 0, 0) and Cl at (½, ½, ½). The symmetry of the space group then generates the remaining six atoms in the unit cell. Another important crystal structure is shown in Figure 16(c), namely that of diamond (and incidentally this is the same for silicon and germanium). This was the second crystal structure to be determined by the Braggs in 1913 and is of especial importance. Once again it is cubic, and with a face-centred lattice. However, to generate the structure, we start with a carbon atom at (0, 0, 0) and a second one at (¼, ¼, ¼) which when combined with the four lattice points in *fcc* gives eight carbon atoms at:

$$(0,0,0),(½,½,0),(½,0,½),(0,½,½)$$
$$(¼,¼,¼),(¾,¾,¼),(¾,¼,¾),(¼,¾,¾)$$

Again, for the record, this would be described using the space group symbol Fd$\overline{3}$m with C at (0, 0, 0). Notice how succinct the space group concept makes the definition of the structure, since only one atom of carbon needs to be specified, and the space group symmetry generates all the other carbon atoms. It can be seen that each carbon atom is at the centre of a tetrahedron of carbon atoms, and it is this arrangement, together with the particular type of bonding between the carbon atoms, that is responsible for the hardness of diamond.

Figure 16(d) is similar to that of diamond, only here the two atoms are different. This is the so-called zinc blende structure, typified by zinc sulfide. The space group is now called F$\bar{4}$3m and one needs to specify a single Zn and a single S atom to generate the complete structure. It is also the structure type of a very important class of semiconductors, as exemplified by gallium arsenide, indium phosphide, etc. Whereas diamond is centrosymmetric, zinc sulfide is not. This results in such a material exhibiting polar properties, such as piezoelectricity: where the application of a stress on the material creates an electric charge (quartz is also a piezoelectric material and is used, for instance, in some cookers to light the gas) or where the application of an electric field causes the material to change shape (used, for instance, to create ultrasonic waves).

The element carbon is capable of adopting a number of different crystal structure types in addition to that of diamond. For instance, graphite (Figure 16(e)) consists of sheets of hexagonal arrays of carbon atoms. This is a completely different type of structural arrangement compared with diamond. The space group symbol is P6$_3$/mmc and only a single carbon position at (0, 0, 0) is needed to enable a crystallographer to draw the graphite crystal structure. This ability of a chemical substance to adopt different crystal structures is called *polymorphism*. Whereas diamond is hard, graphite is soft, and this is why it is used in lead pencils. While the carbon atoms within layers are tightly bonded together, the bonds between the layers are weak and so one layer can slip on another (this makes graphite a good lubricant). However, a single layer is extremely strong, and it is possible to split the graphite layers up to produce the material graphene consisting of a single layer of carbon atoms. This was discovered by Andre Geim (1958–) and Konstantin Novoselov (1974–), who shared the 2010 Nobel Prize in Physics. Another form of carbon called buckminsterfullerene was discovered in 1985, for which Harry Kroto (1939–), Robert Curl (1933–), and Richard Smalley (1943–2005) shared the Nobel Prize in Chemistry in 1996. The molecule has the formula C_{60} and is in the shape of a soccer ball. What a wondrous atom is carbon!

Another important structure is that of perovskite (Figure 16(f)). This is formed by compounds with the general formula ABX_3, where A and B are cations (positively charged ions such as barium or lead) and X is an anion (negatively charged, most commonly oxygen). This class of materials shows many useful properties: electrical, magnetic, piezoelectric, and so on. This arises because of its ability to undergo subtle modifications giving rise to different physical properties.

Figure 16(f) is the high-temperature, centrosymmetric cubic phase of this material; space group $Pm\bar{3}m$, in which the A cation is at the centre of the unit cell, the B cations on the corners, and the anions lie between the B cations to form octahedral environments around the B cations, as seen in Figure 16(f). By making small shifts in the cation positions, a number of different structures and properties can be obtained. In addition, the octahedra can be tilted to form many different structural arrangements: one such tilted arrangement occurs in strontium titanate at low temperature (Figure 16(g)), where the alternating sets of tilts doubles the repeat distance between identical units. The important material $PbZr_xTi_{1-x}O_3$ (or PZT for short), where x can take any value between 0 and 1, is the most widely used piezoelectric material today. There is currently much excitement regarding the use of perovskites, like methylammonium lead iodide, as a new photovoltaic material, as it has a higher efficiency than the conventional silicon cell. However, the explanation for this is so far not clear and is the subject of intense international research as I write.

Another example of an important crystal structure is that of quartz. Quartz crystals consist of SiO_2 groups arranged in a trigonal crystal structure (hexagonal at high temperatures, Figure 16(h)). Each Si atom lies at the centre of a tetrahedron of oxygen atoms, with the tetrahedra joined at their vertices. Glass also consists of SiO_2 groups, but in this case they are not in an ordered array, so glass is not a crystalline material (instead it is an example of an *amorphous* material).

Quartz is a most important industrial crystal: it is part of a multibillion dollar market, where it is used to make precise oscillators, as used in timing devices such as watches and clocks. This makes use of the so-called converse piezoelectric effect, where an oscillating electric field causes a thin slice of quartz to change shape in sympathy with the electric field. By careful choice of the thickness of the slice, it can be made to resonate at a particular frequency, which is then used in the timing.

However, there is a more subtle reason for choosing quartz as an oscillator than simply its piezoelectricity. It turns out that, unlike almost all other piezoelectrics, if one cuts the slice in a very particular orientation (actually there are two possible cuts that can be made) the piezoelectric effect is temperature-compensated. Obviously, if you live in a country with a cool climate, you don't want your watch to gain or lose time when you step off an aircraft in a hot country! This cut has to be made to within a fraction of a degree, and so the quality and expense of your watch depends on how well this has been done.

Organic crystal structures

Crystal structures have also been determined for hundreds of thousands of organic (including metal-organic) compounds. In these crystals, molecules are bonded together in a variety of ways, including the so-called hydrogen-bonding and weak van der Waals bonding. As with elemental crystals, there is still the tendency to pack the molecules together in the crystal structure in order to maximize the density. In Figure 17(a) the arrangement of

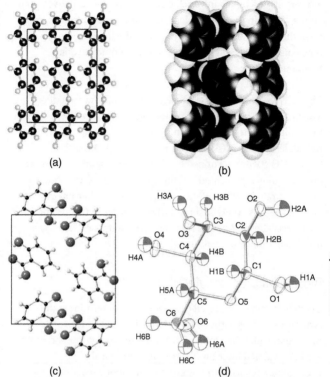

17. Examples of organic compounds: (a). Benzene—black spheres carbon, white spheres hydrogen; (b). benzene space-filling model; (c). aspirin; (d). β-D,L-allose.

molecules of benzene (C_6H_6) at low temperature is drawn using the conventional representation of the atoms as spheres. The unit cell contains four benzene molecules. Drawings like this are often called ball-and-spoke models.

Figure 17(b) shows the same structure, but this time the spheres have been enlarged to sizes compatible with the so-called covalent bonding between the atoms within the molecule (covalent bonds are very strong bonds in which electrons are shared between the atoms). It can be seen how closely packed these molecules are to form the crystal structures. In this case, the bonding between the molecules is weak and hence benzene is a liquid at room temperature.

Figure 17(c) shows a ball-and-spoke model for the painkiller, aspirin. This is acetyl salicylic acid, and you can see how neatly the molecules pack together. Again, this unit cell contains four molecules. Figure 17(d) shows a molecule of β-D,L-allose $(C_{19}H_{29}NO_4S)$ as determined by X-ray crystallography. Here we see the atoms represented by ellipsoids. These indicate how the atoms vibrate in different directions because of thermal motion or possibly because there is some sort of disorder in the atomic positions. These ellipsoids represent parameters called *anisotropic displacement parameters*.

Biological macromolecules

Crystallography has played a major part in determining the structures and activities of large biological molecules, or macromolecules, like DNA, RNA, proteins, and viruses. Proteins are involved in a huge number of functions within living organisms, such as in the catalysis of metabolic reactions, replication of DNA, response to stimuli, and transportation of molecules from one location to another. Proteins are assembled from amino acids using information encoded in genes, and this allows a vast number of different proteins (although note that not all genes in DNA code for proteins). Our bodies produce nearly 10,000 different proteins.

In order to understand how they work in biochemical processes, it is important to know what the molecules actually look like, and this is where crystallographic methods have come to the fore. For example, one of the ways of explaining the action of enzymes is through the so-called *lock and key model*. The way in which a chemical agent (called the substrate) binds to the active site in a specific enzyme relies on the actual shape of the molecule. Often it fits into clefts and holes in the molecular structure. The enzyme molecule is the lock and the substrate is the key. If one wants to design a means of inhibiting an enzyme reaction, it is therefore necessary to have a high-resolution picture of the enzyme molecule.

For the protein or virus crystallographers, the emphasis is on the molecule itself, rather than on the arrangement of the molecules within a crystal, which is incidental. In this case, the main reason for crystallizing biological macromolecules is that the resulting periodic arrangement allows diffraction methods to be used to find the structures of individual molecules.

Proteins are made up of twenty different amino acid residues (small organic molecules containing carbon atoms, and amino and carboxyl groups consisting of a main chain of atoms and a side chain R specific to each amino acid) strung together, rather like beads on a string. Each type of protein is determined by the order in which the amino acids link up, and typically have 100–1,000 individual amino acids. If two of these so-called peptide chains combine together to form a dipeptide, 400 possible combinations result. For a tripeptide there are 8,000 combinations. Clearly, there is an almost infinite number of proteins that can be formed from a polypeptide.

The sequence of twenty amino acids, symbolized by the differently shaded circles in Figure 18(a), constitutes what is known as the *primary structure* of a protein. As was shown by Linus Pauling, Robert Corey (1897–1971), and Herman Branson (1914–1995) in 1951, a *secondary structure* results (Figure 18(b)) from the way the peptide molecule bonds the different amino acid residues together to form the so-called *alpha-helix* arrangement. In proteins, this helix is almost always a right-handed helix (it turns to the right away from you when you look down its length), and is held

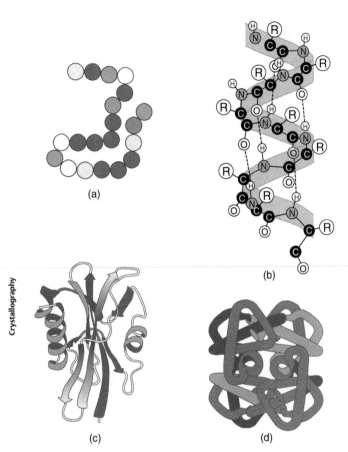

18. **Protein structures: (a). primary; (b). secondary, R marks side groups that determine the particular amino acid residue; (c). tertiary structure of P13-protein; (d). quaternary structure of haemoglobin.**

together by hydrogen bonds. These bonds involve interactions between hydrogen atoms and nearby oxygens (marked by the dotted lines). Another secondary structure, the beta sheet, is formed when the peptide chains fold to lie more or less parallel or antiparallel to one another.

Figure 18c shows an example of a tertiary protein structure in which the protein has now been folded up to give a particular shape to a protein macromolecule. The coiled ribbons and flat arrows indicate the alpha-helices and beta-sheets, respectively. It is this shape that is critical in affecting the way the protein functions, and this is why it is important to use a technique like X-ray diffraction to determine the folding of the various helices and sheets. Certain disorders such as Alzheimer's, Parkinson's, and type 2 diabetes result from the misfolding of proteins. Figure 18(d) illustrates an example of a quaternary protein structure, in this case of haemoglobin, where two or more polypeptide subunits are connected together. Haemoglobin consists of two alpha and two beta chains, with 141 and 146 amino acid residues per chain, respectively. Many enzymes also form quaternary structures.

In order to find the true structure of a particular protein using X-ray diffraction, it is first necessary to crystallize it. This is often the rate-limiting step in protein structure determination, since it is impossible to predict the best growing conditions. These days, when it is necessary to carry out many hundreds, or even thousands, of such determinations, this process requires a certain amount of automation.

Crystallographers working on biological macromolecules now have robots that pipette the protein plus different solvents into a series of plastic cells in order to find which ones grow the best crystals. Such is the sophistication that there are now robots that are able to place the crystal on to a diffractometer in the correct position for the X-ray beam, often at a synchrotron source. The first structure determination of an enzyme, lysozyme, in 1965 took many months of data collection and subsequent analysis. Such is the progress made in recent times that today this whole operation can be done within an hour or so.

The fact that proteins are chiral molecules means that the space groups defining the crystal symmetry cannot be those that include reflection planes or centres of inversion, as these symmetry operations would have the effect of changing the handedness of the molecules. This leaves sixty-five possible space groups out of the total of 230. In addition, it has been found that roughly one-third of all protein crystal structures belong in one particular space group, denoted $P2_12_12_1$. The letter P means that the lattice is primitive and the three 2_1 symbols mean 2-fold screw axes directed along three mutually perpendicular crystal axes. The crystal system is orthorhombic. Several papers have tried to explain this phenomenon.

Over the years, crystallographers have determined protein and related structures of increasing complexity. For instance, in 2009 the Nobel Prize was awarded to Venki Ramakrishnan (UK), Tom Steitz (US), and Ada Yonath (Israel) for their studies of the structure and function of the ribosome. The ribosome acts within the cell as a kind of 'machine' that produces proteins. The sequence of DNA encoding for a protein may be copied many times into RNA chains of a similar sequence. The ribosomes can then bind to an RNA chain and use it as a template to make the correct sequence of amino acids in a particular protein. The ribosome is a real monster, requiring in excess of 293,000 atoms to be located: it has a diameter between 200 and 300 Å. The solution of such a structure, which showed how the ribosome does its work, was an awe-inspiring tour de force.

In addition to protein structures, it is also possible to use crystallographic techniques to study the structures of virus particles. Viruses consist of a genome of DNA or RNA usually surrounded by a protective coating (called a capsid) of proteins or lipids. Crystallography is a valuable tool of drug and vaccine discovery, and this has led to the development of antiviral compounds, especially against HIV enzymes. To do this, as with proteins, it is first necessary to get the virus particles to crystallize together, and then employ similar methods as used for proteins.

Viruses adopt different shapes, although many have an almost spherical form. For instance the foot-and-mouth disease virus (FMDV) has a high degree of icosahedral symmetry and its genome consists of a single strand of RNA: it crystallizes into a body-centred cubic array. In 1989, a near-atomic resolution determination of this virus was made by Ravindra Acharya, Elizabeth Fry, David Stuart, Graham Fox, David Rowlands, and Fred Brown, using data collected with synchrotron radiation. The cubic unit cell has a volume of 41,064 nm³ (228,862 times as big as that of sodium chloride!) and contains just two virus particles.

The poliovirus has a similar shape and there is current research into preparing a polio vaccine based on knowledge of the virus structure. The present vaccines use a weakened version of the poliovirus to stimulate the immune system. However, in some cases a reaction is set up in the gut, and this allows a reactivated virus to pass out of the body and spread to other, unvaccinated people. There is now research into the crystallography of this virus with the aim of creating a synthetic virus that contains no genome, which means that it cannot replicate, but may, nonetheless, trigger the appropriate defensive result in the body. This has already been achieved for one strain of FMDV in 2013.

Non-ambient crystallography

In addition to determining crystal structures, crystallographers often want to study how these structures change when, for example, temperature or pressure is altered. There are many reasons for wanting to do this, such as researching into the connection between structure and physical properties. Suppose on heating or cooling a material changes its crystal structure, thus undergoing what is called a *phase transition*. When this happens, certain properties may undergo a sudden change and so by studying structure and properties at the same time one can find a link between them. For example, when one heats a crystal of barium titanate, it undergoes a massive change in its dielectric behaviour, and by studying this we can relate the effect to small changes in the displacements of barium and titanium atoms in the structure.

Application of pressure is also important since it often causes a crystal to adopt completely new and unimagined structures. One area of considerable interest is in understanding the nature of minerals deep within the earth, and so application of very high pressure and sometimes simultaneous heating can reveal what is happening at great depths underground. An interesting example is work carried out in Japan a few years ago into what happens when metals such as iron are subject to intensely high pressures and temperatures. It was found that under these conditions large crystals formed, and it was suggested that this may mean that in the earth's core there are crystals several kilometres in length, and that this has implications for understanding terrestrial magnetism (see Preface).

There are particular reasons for wanting to collect diffraction data at a low temperature. First of all, lowering the temperature causes atoms to vibrate less, and this has the effect of increasing the intensities of reflections at high angles. Thus more data are measured, and the resulting crystal structures are determined to higher resolution. Therefore, if you look at published crystal structures in the scientific journals, it is highly likely that they will have been carried out typically at about 100 K (−173°C).

Another reason to cool crystals is to solve a problem that used to limit the ability to obtain high-resolution diffraction information from biological samples. The difficulty here was that protein and virus crystals tend to be destroyed fairly quickly when placed in the X-ray beam. The result was that in order to obtain complete information, data collection had to be done on several crystals, and then the various datasets had to be combined on the same scale. In 1970 there was a breakthrough, when it was discovered that if a protein crystal was flash-cooled and maintained at a low temperature it survived for much longer (now known to be around a factor of seventy times) in the beam. In the 1980s this led to a new discipline within crystallography called *cryocrystallography*, and this cooling is now standard procedure for protein diffraction studies, whether in the laboratory or at synchrotron sources. Many of the recent Nobel Prizes, such as for the study of the ribosome, have used cryocrystallographic techniques.

Many different methods for cooling crystals were developed in the period since the discovery of X-ray diffraction. Because the X-rays used in structural studies tend to be absorbed by materials placed in their path, the usual method has been to use an open stream of a cold gas, such as nitrogen, blowing over the crystal. Early apparatuses for doing this were crude, very unstable, used large quantities of liquid nitrogen in order to generate cold gas, and in general were clumsy to use. Kathleen Lonsdale used to cool crystals by simply dropping liquid air on to a crystal. I like the following graphic description given by the famous metallurgist William Hume-Rothery (1899–1968), who remembered a visit to the Royal Institution where he saw 'the figure of Dr. Lonsdale appearing through a cloud of mist, like a glorified spectre of the Brocken, while her assistant pumped liquid air over a crystal'.

A consequence of the difficulty of cooling crystals with an open stream of gas was that most crystallographers avoided crystal cooling altogether. However, around the same time that flash cooling was discovered in protein crystallography, a new device, known as the Cryostream was invented by John Cosier and myself.

The basic principle of the Cryostream is to have an open Dewar vessel containing liquid nitrogen, and then suck the liquid out into a heat exchanger. Here the liquid nitrogen is vaporized by a heater, and exits the heat exchanger through a pump, where the gas is warmed to near room temperature. It is then passed back to the heat exchanger, whence it is recooled by the cold liquid on the other side of the heat exchanger. The resulting cold gas then exits through a nozzle on to the crystal. In the nozzle is a computer-controlled heating element enabling the operator to set the temperature of the nitrogen stream anywhere between 90 and 350 K, with a temperature stability of ± 0.1 K. Because there is no pressure difference between the Dewar and output flow, the system can be refilled at any time with no effect on the outlet temperature. The temperature stability and use of flexible hoses to aid in lining up the cold stream on the diffractometer was of particular importance. The Cryostream has been so successful that today it is in use in almost all the crystallography laboratories and synchrotron establishments worldwide, and low-temperature crystallography is now routine.

In the last thirty or so years, great advances have also been made in techniques for applying high pressures to crystals. This is usually achieved by the diamond anvil method. Here, a tiny crystal is placed within a small hole in a thin metal gasket. Into this hole a liquid, for example a mixture of methanol and ethanol, is added, and sometimes a tiny chip of ruby is included. This is then sandwiched between two diamonds whose points press into the gasket hole thus creating a sealed chamber. When the diamonds are forced together the pressure in the liquid is increased hydrostatically, and this pressure is transmitted to the crystal specimen. The whole apparatus is placed on a diffractometer, and is aligned so that the X-ray beam passes through the diamonds at the same time as the crystal of interest. The ruby chip is used to measure the actual pressure by using light from a laser, which stimulates fluorescence. The wavelength of the fluorescence varies with pressure, and this can therefore be used to calibrate the pressures in the diamond anvil cell. With this apparatus immense pressures can be achieved.

Growth of crystals

In the last one hundred years or so, many techniques have been developed in order to grow single crystals of varying sizes. The basic principle of crystal growth is to start with a small 'seed' crystal, or even sometimes a speck of dust will do, and then let the molecules or atoms accumulate around the seed. Provided conditions are adjusted correctly it is then even possible to grow nearly perfect, large crystals.

Solution growth, in which the substance of interest is first dissolved in a solvent in sufficient quantity as to nearly saturate the solution, is perhaps the simplest of the techniques. Crystal growth then proceeds by evaporation of the solution or by change of temperature, usually by cooling. In either case, the concentration of the substance is increased, and it eventually starts to precipitate out. If this occurs quickly, then usually very small crystals are formed. On the other hand, if the temperature is controlled carefully, and especially if a seed crystal is dangled in the solution, it is then possible to grow a single extremely large crystal. International Crystallographic Associations sometimes run crystal growing competitions in schools using this method, often with the chemical potassium alum. It is relatively easy to grow beautiful octahedral crystals measuring several centimetres across.

When chemists synthesize a new compound they will often precipitate it out in solid form from a solution. The standard method that all chemistry students learn is to scratch the inside of the test tube with a glass rod: this creates numerous microscopic glass 'seeds', thus forcing precipitation. For chemists, this is usually used as a means of purifying their product. Precipitation by seeds, often dust or pollen, is also the cause of snowflakes forming in the clouds.

For most organic crystals, the solvents used are liquids at room temperature, typically water, acetone, alcohol, and other organic solvents. It is also possible to use high-temperature solvents, especially for growing inorganic crystals. In this case one uses as the solvent a chemical that melts at a high temperature, which then dissolves the material of interest. This is known as *flux growth*. For example, for many years in the 1950s crystals of the industrially important material, barium titanate (used, for example, in electrical capacitors), were grown by mixing barium titanate powder with an excess of potassium fluoride placed in a platinum crucible. This was then heated in a furnace until the fluoride melted and dissolved the barium titanate. It was held at high temperature for several hours and then slowly cooled until crystals of the titanate formed on the walls of the crucible. The hot solvent was then poured off, and the crystals extracted.

In this way, quite large crystals were produced, and indeed for many years most published scientific research into this material used such flux-grown crystals. However, it has to be said that the crystals contained impurities, mainly of fluoride ions from the flux, and this had a number of effects on the properties of barium titanate. For instance the flux-grown crystals underwent a phase transition from a tetragonal to a cubic structure at 120°C; but several years later when new methods enabled barium titanate crystals of higher purity to be made, the transition temperature increased to nearly 140°C, and the pure crystals had better optical properties. Even the shapes of the crystals were different: plate-like for flux-grown crystals, more block-shaped for the purer crystals.

The usual method for crystal growth, especially of materials for which solvents are not available, is to melt the material and then allow it to solidify under controlled conditions to form a single crystal. Many different methods have been devised in order to do this, the most common one being the *Czochralski method*. This is named after the Polish scientist Jan Czochralski (1885–1953) who discovered the technique, as so often happens in science, by accident.

Czochralski was born in a small town called Kcynia near the city of Poznan, which at the time was under Prussian rule. This meant that, despite being a Pole, officially he had German citizenship. His early history is rather vague, even to the point that we do not know the nature of his qualifications for scientific research. Nonetheless, he went to Germany and carried out research in the metals industry. The story goes that one day, while working at the laboratory bench, he dipped his pen nib into what he thought was an ink-well; however, it was a pot containing molten tin. He noticed that when he withdrew the pen, a long sliver of metal remained attached. Now, many people would simply have discarded the pen and got on with some other work, but Czochralski had the wit to investigate this effect. As was once said by Pasteur, 'Dans les champs de l'observation le hasard ne favorise que les esprits préparés' ('In the field of observation, chance favours the prepared mind').

Czochralski had already begun to use X-ray diffraction in his work, and soon discovered that the sliver of metal was a pure crystal of tin. He experimented with different rates of pulling the pen out, and he found that larger crystals could be grown by slow pulling. He probably was not fully aware of the importance of his discovery, but it was many years later that germanium, and later silicon, crystals were grown by Gordon Kidd Teal (1907–2003) working at Bell Labs, US. His method was an adaptation of Czochralski's method, but he failed to fully acknowledge this. Czochralski never made any money from his invention and, before World War II, he went to Warsaw by invitation of the President of Poland to take up a professorship at the Warsaw Institute of Technology. However, after the war he was accused of not being a true Pole, and of collaborating with the Nazi occupiers. He narrowly escaped imprisonment; his name was removed from the list of professors at the university, and he died in ignominy in his home town of Kcynia.

Recently, though, it has been discovered from the release of secret files that, in fact, he was secretly working for the Polish resistance home army during the war by making use of his German contacts, and was probably even working for the Polish secret service earlier while in Germany. His reputation has now been fully restored, including his name on the professorial list, and in Poland today he is highly revered for his scientific work.

Today the Czochralski technique is used to grow huge crystals of silicon, measuring as much as 300 mm across. Crystal boules pulled in this way tend to be cylindrical in shape. This is usually done by melting the material in a crucible and then suspending a small seed crystal on a fibre in the molten liquid. Motors then pull the fibre out very slowly while at the same time rotating the growing boule about the fibre axis. From these boules large thin wafers are cut, to be used in the microchip market.

Several other melt-growth methods have been invented. For instance, the *Bridgman–Stockbarger method* involves heating a polycrystalline material above its melting point within a sealed ampoule, and then slowly pulling the ampoule through a temperature gradient. This is cooled from one end where a seed crystal is located. A single crystal of the same crystallographic orientation as the seed material is grown on the seed and is progressively formed along the length of the container. The process can be carried out in horizontal or vertical geometry. The Bridgman–Stockbarger method is a popular way of producing certain semiconductor crystals such as gallium arsenide, for which the Czochralski process is more difficult.

Another method, due to the French chemist Auguste Verneuil (1856–1913) who invented it in 1902, involves melting a finely powdered substance using an oxyhydrogen flame, and allowing the melted droplets to fall on to a rod on which the crystal then grows. This has been used to good effect to make artificial sapphire and ruby crystals, as well as strontium titanate.

A more modern method is the *floating zone method*, where a liquid zone made by melting is moved slowly through the material. This has the effect of purifying the solid, and if seeded properly results in the growth of a single crystal. The latest

19. Top: mirror furnace. Bottom: example of a crystal of CoSi$_2$O$_4$; A is the polycrystalline feed material, B is the melt zone, and C is the crystal.

apparatus for this uses an intense light source and curved mirrors to focus the light on to the melt zone. In the example shown in Figure 19, a maximum temperature of 2,000°C is reached using four 1.5 kW halogen bulbs. A feed rod of the material is lowered through this hot zone and below this zone the solid crystallizes. This technique is capable of growing a large single crystal within a few hours. Because the sample material is not in touch with any crucible or other substance the resulting crystal is of high purity.

Crystals of quartz are vitally important for their piezoelectric properties, especially for their use in the electronics industries as oscillators and timing devices. This is a huge crystal industry. These crystals are grown by the *hydrothermal method*, which uses the fact that the solubility of quartz crystal in an alkaline liquid varies with high temperatures and pressures. The vessel for growth is a pressure vessel called an autoclave, and in industrial processes may be several metres in height. Raw material, called lasca (small pieces of natural quartz crystals), is placed in the lower part of the autoclave and in the upper part, many seed crystals of quartz in the form of thin plates are suspended in racks above a permeable baffle. The vessel is filled with water and a mineralizer such as Na_2CO_3 or NaOH, and after encapsulating, it is heated up to very high pressure (this is a dangerous method of crystal growth!). Finally, it reaches a supercritical state, and the growth begins. By keeping the temperature of the upper part of the autoclave below that of the lower part, convection occurs and the solution is moved from the lower part to the upper part, where quartz crystals precipitate on to the seed crystals. Since the synthetic quartz crystals are grown under strict control, they have precise shapes, dimensions, and properties. Almost perfect crystals of quartz, several centimetres in length, can be grown like this over a period of weeks or months.

Controlled vapour deposition (CVD) is a popular industrial method to grow crystals by depositing the solid material from a vapour on to a cold surface. There are many variants, which have been used to make semiconductor thin films. It is also used to make artificial diamonds. While diamond growth can be done by applying high pressures and temperatures to carbon in a press, CVD has increasingly been employed instead. In this technique a hydrocarbon gas is passed into a chamber on to a substrate, and diamond crystals form by chemical reactions that are still a matter of research. CVD is useful because it enables diamond crystal to be deposited over large areas, so that one could, for instance, coat electronic components. This is of interest because diamond has the unusual property of being a good electrical insulator but at the same time an excellent thermal conductor, thus enabling heat to be carried away from the electronic components. Today, gem-quality diamonds can be grown artificially, and in order to protect the diamond gem market, spectroscopic and other methods have been developed in order to distinguish between synthetic and natural diamonds.

None of these methods are suitable for the growth of protein crystals, for which special techniques have been developed. The most commonly used are hanging and sitting drop vapour diffusion methods. The hanging-drop method involves a drop of protein solution being placed on an inverted cover slip, which is then suspended above a reservoir containing a liquid reagent. The sitting drop crystallization apparatus places the drop on a pedestal that is separated from the reservoir.

Chapter 4
Diffraction

Reciprocal lattice

When electromagnetic radiation (visible light, X-rays, etc.) passes through small apertures with a size comparable to the wavelength of the radiation, it has long been observed that the beam of radiation becomes scattered. This is called *diffraction*. While it is relatively easy to deal with diffraction from a one-dimensional array of objects, in order to understand diffraction from a three-dimensional crystal it is convenient to define a new type of lattice constructed from the real lattice that we have used so far. This is the so-called *reciprocal lattice*, proposed by Paul Ewald around 1911.

We begin by considering how this lattice is constructed, and then we shall investigate its uses in understanding diffraction. Figure 20 shows how to construct it for an example of a two-dimensional lattice. In Figure 20(a) we start with a unit cell with axes a and b, here chosen deliberately to be non-orthogonal. We start by constructing a line (dashed) through the corner of the unit cell and perpendicular to the (100) planes of spacing d_{100}. Along this line mark a point, labelled 100, at a distance $1/d_{100}$ (note that physicists usually multiply by a factor of 2π, whereas crystallographers use simply 1). Twice as far away mark a point called 200, corresponding to (100) type planes of half of d_{100}. This can be continued to place

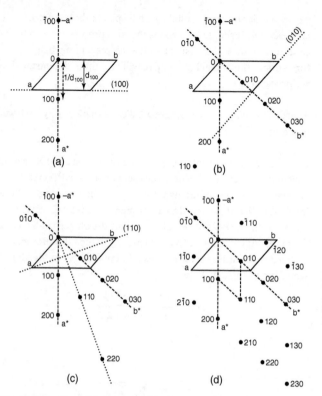

20. Construction of the reciprocal lattice.

300, 400, etc. In the opposite direction, we mark the points $\bar{1}00$, $\bar{2}00$, etc. This gives a reciprocal lattice axis a*.

Next choose the (010) type planes as in Figure 20(b), to obtain the series of points 010, 020, 030, $0\bar{1}0$, $0\bar{2}0$, etc., and the axis b*. In Figure 20(c), the (110) plane is drawn and again a line is drawn perpendicular with points placed at distances inverse to the interplanar spacing d_{110}. Finally, in Figure 20(d) the points corresponding to other planes have been added. The result is a new lattice where each point arises at the end of a vector perpendicular to a real crystallographic plane: the *reciprocal lattice vector*. It should be obvious that this process can also be continued in three dimensions, so that corresponding to a three-dimensional real lattice there will be a three-dimensional reciprocal lattice.

Apart from being a nice mathematical construction, why have we bothered to do this?

In Figure 21, C represents a crystal, and AC is a beam of X-rays (or neutrons or electrons) incident on a set of planes in the crystal at an angle θ. A sphere of radius $1/\lambda$ where λ is the wavelength of the incident radiation, is now drawn through the crystal C. This is known as the *Ewald Sphere*. CP is a vector drawn perpendicular to the planes. If CP happens to be of length $1/d$, then it represents the reciprocal lattice vector for this set of planes. It can then be seen from triangle ACP that

$$\sin\theta = \frac{CP}{AC} = \frac{1/d}{1/\lambda} = \frac{\lambda}{2d}$$

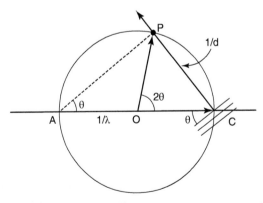

21. **The Ewald Sphere construction.**

which should be familiar as Bragg's Law (if not, see Chapter 1). In other words, if a reciprocal lattice point lies on the surface of the Ewald Sphere, Bragg's Law is obeyed and a diffracted beam will be obtained along the direction OP, drawn from the centre of the Ewald Sphere through the reciprocal lattice point. If the orientation of the crystal planes is such that the corresponding reciprocal lattice vector places the lattice point away from the surface of the Ewald Sphere, then no diffraction is possible from that set of planes.

This clever construction now illustrates an important point about crystal diffraction. Suppose that Laue had been correct in his belief that the X-rays he used consisted only of a single wavelength, or even five wavelengths. Then the chances of having the crystal in the correct orientation to enable several reciprocal lattice points to lie on the Ewald Sphere at the same time would have been remote, and the discovery of X-ray diffraction would not have happened at that time. WHB would not have taken an interest in X-ray diffraction, and WLB would not have begun the field of X-ray crystallography. Laue was lucky that the X-ray beam consisted of a continuum of wavelengths: one can think of this in terms of having Ewald Spheres with radii between very small and very large (Figure 22(a)) values. Then, with a stationary crystal all those reciprocal lattice points in the greyed area can be seen on the film—the spots arising from different wavelengths satisfying Bragg's Law. This type of diffraction is known as *Laue diffraction* (Figure 22(b)).

If, on the other hand, one uses a monochromatic beam, it is then necessary to rotate the crystal, and hence the reciprocal lattice, so that at some time during the rotation each reciprocal lattice point passes through the surface of the Ewald Sphere, giving rise to a reflection. For many years, this was achieved by a variety of ingenious X-ray cameras designed to rotate or oscillate the crystal, sometimes with a simultaneously moving film to spread the reflections out. This meant that the crystal had first to be carefully oriented, and a number of clever techniques were devised to do this; some of them were passed down from graduate supervisor to

(a)

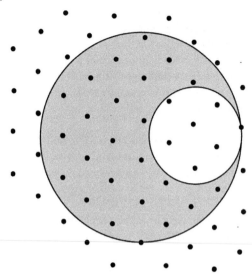

(b)

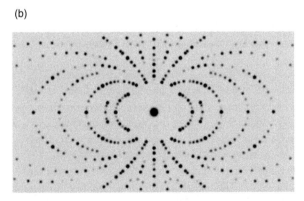

22(a). Formation of a Laue pattern by a continuum of wavelengths using the Ewald construction; (b). example of a Laue photograph.

student without ever being published. When I started crystallography research in 1965, the intensities of the reflections were usually measured by eye, through simply comparing each spot with a specially prepared scale, typically in a range of 1 to 10. To measure a full set of data involving perhaps 1,000 reflections took several days' painstaking work. It is remarkable that with such crude measurements it was possible to work out the structures of quite complex crystals.

However, these days photographic methods are hardly used, and indeed it is now difficult even to purchase suitable X-ray films. Instead, for the study of single crystals, modern diffractometers (Figure 7) are used in almost all crystallography laboratories, the latest ones having area detectors based on CCDs, or more recently CMOS, for the study of small-molecule crystals (similar to the imaging devices in mobile phones, but much larger). Larger silicon pixel detectors are used in macromolecular crystallography. The increase in computing power now available enables many thousands of reflections to be accurately measured, and crystal structure determination to be made, that was impossible to imagine only twenty or so years ago.

The crystallographer first selects a suitable crystal, usually under a microscope, typically 0.1–1 mm in size (usually even much smaller for protein crystals), and this is stuck with glue to a thin glass fibre, or in the case of biological crystals, where crystal cooling is usually employed, mounted in a tiny loop. A common method used for transferring biological crystals to the diffractometer is by first covering the sample with an inert cold oil after selection of a suitable sample under a microscope. The sample is mounted on a special device called a *goniometer head*, and this is then placed on to the diffractometer. The crystal is centred with the use of a telescope so that it can rotate or oscillate within the X-ray beam. Today's diffractometers are fully computer-controlled so that it is no longer necessary to orient the crystal, but instead software sorts it out for you.

A diffractometer fitted with an area detector collects many two-dimensional images of data, all of which are then combined. With synchrotrons it can take as little as five minutes to collect a complete dataset from something as complex as a protein. The software works out the probable unit cell, crystal symmetry, and the indices of each spot. Whereas data used to take many hours or days (or even weeks) to collect for any crystal, today this is often accomplished automatically in an hour or two. The control software actually makes use of the reciprocal lattice and the Ewald construction behind the scenes to drive the detector to various angles and rotate the crystal into suitable positions. Today, crystallographic software is so sophisticated that after data collection the crystal structure can often be determined automatically, virtually without human interaction. Nevertheless, especially in protein structure solution, a model for the molecule has still to be built and checked using computer graphics by a human being.

However, this is not true for all cases, and one still has to be careful that any automated process has not led to a false result. The literature contains many examples of incorrect structure determinations. In reality, one has to be aware that the crystal might, for example, be twinned, i.e. two or more parts of the crystal may have grown together in different orientations, so that the diffraction pattern observed is actually a superposition of separate patterns. This complicates the analysis of the data, but can generally be sorted out by a competent crystallographer. Unfortunately, the advent of push-button automation has tended to mean that many people without good training in crystallography can use modern diffractometers, but I fear that their lack of experience may mislead them.

Reflection intensities and amplitudes

Each wave arriving at the detector after diffraction by the crystal can be described by its amplitude and its relative phase difference

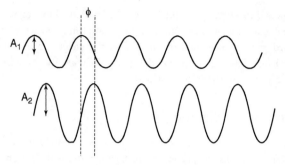

23. Example of two waves with different amplitudes and with a phase difference ϕ.

(Figure 23). However, what is recorded at the detector is the intensities of the diffracted beams, which are proportional to the square of the amplitudes. In measuring a quantity proportional to the square of the amplitude, all relative phase information is lost. Crystallographers write the amplitude in terms of the so-called *structure factor F(hkl)*, which represents the amplitude of scattering from the set of crystal planes given by the Miller indices h, k, and l. The intensity for any reflection is therefore given by

$$I(hkl) \propto |F(hkl)|^2$$

It turns out that the diffraction pattern appears to be centrosymmetric as it can be shown that

$$I(hkl) = I(\bar{h}\bar{k}\bar{l})$$

This relationship is called Friedel's Law, and applies provided that the X-rays are not being absorbed by an atom in the crystal. The X-ray intensity typically shows an absorption peak at a particular wavelength. If the wavelength is such that absorption by an atom in the crystal does occur, this law breaks down: this effect is known as *anomalous dispersion*. Usually the anomalous effect is weak and to a first approximation can be ignored. However, it is possible to tune the wavelength of the X-rays using a synchrotron in order to enhance the anomalous scattering effect. A breakdown of Friedel's Law can be used to determine the chirality or polarity of a crystal and of its atomic structure.

Convolution theorem

A neat way to understand the intensities seen in diffraction patterns obtained from a crystal is to return to the definition of a crystal structure by convolution, which for an infinite crystal is

$$C = M * L.$$

As a reminder, C stands for the crystal structure, M for a molecule, and L for the lattice. We now need to appeal to a mathematical quantity called the *Fourier Transform*. Without going into the mathematical details, this is a way to calculate the amplitude of a diffracted wave from the diffracting object, in our case the crystal. In addition, there is a special mathematical tool called the *convolution theorem*, which says:

> The Fourier Transform of the *convolution* of two functions is equal to the *product* of the Fourier Transforms of each function, and vice versa.

Therefore, the diffraction amplitudes from the crystal are given symbolically by

$$C(FT) = M(\text{FT}) \times L(\text{FT}),$$

where FT refers to Fourier Transformation. To understand this we need now to consider the terms $M(\text{FT})$ and $L(\text{FT})$ separately and then multiply them. Fortunately, multiplication is a much simpler mathematical operation than convolution.

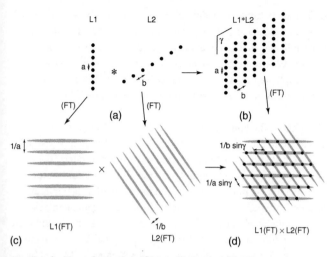

24. Fourier Transformation of a two-dimensional lattice.

A useful way to think of Fourier Transformation is through a couple of simple 'rules', namely 'symmetry is preserved' and 'lengths are inverted'. To illustrate this, consider the two-dimensional lattice array of points in Figure 24.

First of all, start with two one-dimensional lattices denoted L1 and L2 (Figure 24(a)). The convolution L1 * L2 then gives the two-dimensional lattice in Figure 24(b). If we now apply my Fourier Transform rules to L1, the translational symmetry is preserved in the vertical direction, except that the repeat distance is inverted. At the same time, because the extent of L1 in the horizontal plane is essentially infinitely small, in the Fourier Transform it becomes infinitely extended as suggested by L1(FT) in Figure 24(c). Similarly, the Fourier Transform of L2 looks like L2(FT).

Finally, in Figure 24(d) we find the product L1(FT) × L2(FT). Because the two functions are multiplied together they only have values different from 0 where they cross, giving rise to lines perpendicular to the two-dimensional plane. In other words, the diffraction pattern that would be expected from a two-dimensional crystal would consist of a regular array of long *rods* of scattering perpendicular to the crystal plane (projected as dark spots in the Figure). Clearly, in the case of a three-dimensional lattice the effect of multiplying L1(FT) × L2(FT) × L3(FT) would give rise to a regular array of spots, corresponding to the reciprocal lattice.

What about the Fourier Transform M(FT)? If M consists of a single atom, then the Fourier Transform is that from a very small sphere in shape. Thus the diffracted waves form a spherical distribution (rule 1). On a photographic film, this would simply look like diffuse intensity peaking in the forward direction and gradually weakening in intensity as the scattering angle increases (this would be what the diffraction from a gas of atoms would look like). In addition, the larger the atom is the smaller the diffracted size (rule 2). Thus, in the case of X-rays, which are diffracted by the ball of electrons around an atom, the more electrons, and hence greater size, the quicker the diffracted amplitude falls off with angle.

Another way to visualize this is that the X-rays in the forward direction are scattered *in phase* by all the electrons in the atom, thus making the amplitude proportional to the atomic number (the number of electrons in the atom). However, on scattering to angles away from the forward direction, the waves scattered by individual electrons undergo interference, thus reducing the amplitude of scattering as the angle increases. Tables of atomic X-ray scattering factors are available for calculation purposes. It should be obvious that since the amplitude is proportional to atomic number, light atoms scatter X-rays less than heavy ones.

Suppose that the crystal consists of repeating molecules rather than individual atoms. In Figure 25(a) we start with a molecule M: in this case it is a simple hexagonal ring of atoms. Figure 25(b) shows its Fourier Transform M(FT): note how this has preserved the 6-fold symmetry of the molecule.

In Figure 25(c) there are two molecules, which we can think of as a convolution of one molecule M with a finite horizontal 'lattice' L consisting of only two points. From the convolution theorem we then expect to see the Fourier Transform of the molecule *multiplied* by the Fourier Transform of this two-point lattice. This gives rise to the image in Figure 25(d), where we again see the underlying Fourier Transform of the original molecule. But this time it is crossed by a set of vertical fringes, with spacing inversely proportional to the distance between the two original lattice points of L.

In the next case (Figure 25(e)), the molecule has been convoluted with a two-dimensional lattice and the Fourier Transform (Figure 25(f)) again reveals the underlying transform of the molecule, but this time crossed by vertical and horizontal fringes to give in projection a repeating pattern of spots of different intensities arranged in a regular array (the reciprocal lattice). Finally, Figure 25(g) shows what happens if the molecules are further apart. This causes the fringes to become close together and this time we can see even more clearly the underlying Fourier Transform of the molecule (Figure 25(h)). It is clear from this that the pattern of intensities in the diffraction pattern reveals information on the symmetry of the diffracting object, i.e. the molecule. It is the job of the crystallographer, when faced with a diffraction pattern, to use these intensities in order to locate the atoms in the original molecule and decipher how they are stacked together in the unit cell.

Powder diffraction

When radiation is diffracted by a powdered material, the result is a series of rings each of a constant intensity. Figure 26 shows how this occurs. First of all, it should be understood that a polycrystalline powder consists of a random collection of very small crystals (crystallites), typically of the order of 0.1 mm or less. In Figure 26(a), the diffraction spots from a single crystal are

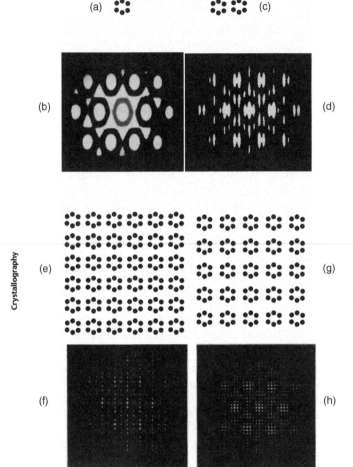

25. Formation of a crystal diffraction pattern using a molecule and a lattice.

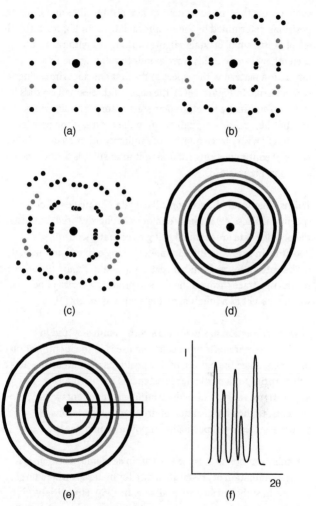

(a)

(b)

(c)

(d)

(e)

(f)

26. Formation of powder diffraction rings.

shown: note that the spots have different intensities and lie on the positions determined by the reciprocal lattice. In Figure 26(b), the effect of two small crystals, slightly misaligned with respect to each other is shown: this corresponds to rotating the first diffraction pattern with respect to the first and superimposing the two patterns. In Figure 26(c), the effect of three small crystals is shown. You can see that the spots start to form annular rings around the straight-through beam in the centre. If the powder consists of a very large number of randomly oriented crystallites, a series of continuous rings of different intensities is formed, as shown in Figure 26(d).

This is what would be seen on a flat-plate film placed after the specimen. These days, film techniques are hardly used, but instead the powder data are scanned using a computer-controlled detector (in a powder diffractometer) along a line from the centre of the pattern towards higher scattering angle (Figure 26(e)). This results in a trace through the lines to produce a series of peaks of varying height as a function of angle 2θ (Figure 26(f)).

Powder diffraction has many uses both in industry and in academic research. Its most common use is in the identification of materials, since the pattern of line positions and intensities acts as a fingerprint for a particular substance. This has many applications, ranging from determining components in, say, cements, to the identification of drugs and polymorphs of pharmaceuticals—of particular importance in patents.

Another common use, especially within academic research, is in the determination of crystal structure for materials where no single crystals of sufficient size are available. In 1969, Hugo Rietveld (1932–) showed that if one starts with a reasonable model for the structure, it is possible to refine the atomic coordinates by fitting the intensities of the whole pattern, including parameters specific to the peak shapes and other geometrical factors. This so-called Rietveld refinement process is now one of the most common techniques for obtaining structural information from powdered specimens. Furthermore, considerable progress has been made in solving crystal structures from powders even when no initial model is available, especially for high-symmetry cases. These topics are today so important that conferences are held regularly to discuss the latest advances in powder diffraction.

Incommensurate (modulated) crystals

So far, we have considered crystals described by periodic translational symmetry. However, in reality crystals are rarely perfectly periodic, and over the last hundred years departures from periodicity have been found. It was known for some time that crystals of calaverite (an alloy of gold and tellurium, and sometimes silver) were decidedly odd. It was noticed as far back as the turn of the 20th century that the crystals seemed to show ninety or more different forms, and that to index the faces of the crystals seemed to require Miller indices of unusually high values. Later, it was discovered that this could be improved by using four indices instead of the usual three.

In 1927, U. Dehlinger explained the appearance of extra spots in certain diffraction patterns in terms of periodically arranged defects. Then in 1940 a curious phenomenon was found by Albert James Bradley (1899–1972), while working on X-ray diffraction from copper-iron-nickel alloys in the laboratory of WLB. He observed sharp powder diffraction lines as expected, but in addition he noted that each line was flanked by two slightly diffuse, but quite strong, bands. Initially he interpreted this as resulting from a mixture of two different phases of the material. However, Vera Daniel (1917–1993) and Henry Lipson (1910–1991), working in the same laboratory, showed in 1943 that this effect was explicable in terms of a basic structure that is modulated by some periodic variation whose wavelength is much larger than the length of the unit cell. Such a modulation was then incommensurate with the fundamental unit cell dimension.

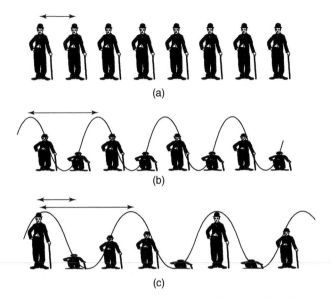

(a)

(b)

(c)

27. Commensurate and incommensurate modulations: (a). regular periodicity; (b). commensurate modulation; (c). incommensurate modulation.

Figure 27 illustrates the idea of modulation in a one-dimensional crystal. In Figure 27(a), we see a periodic array of molecules. In Figure 27(b), a wave of wavelength twice the repeat distance of that in 27(a) is superimposed to create a periodic alternating array of large and small molecules. In this case the true repeat of this array has been doubled, and is therefore an example of a commensurate modulation. However, in Figure 27(c) the wave has a wavelength that is out of step with the fundamental spacings of the molecules, so that the resulting structure no longer has a simple repeating periodicity. This is an example of an incommensurate modulation.

Little else was heard of incommensurate structures until 1964, when the Dutch crystallographer Pim de Wolff (1919–1998), working in Delft, found a similar anomaly in crystals of dehydrated sodium carbonate. He showed that the diffraction pattern could not be indexed solely by three integers, but needed four. Thus the reciprocal lattice vector H was written with four indices h, k, l, and m, thus

$$H = h\mathbf{a}^* + k\mathbf{b}^* + l\mathbf{c}^* + m\mathbf{q}^*,$$

where $\mathbf{q}^* = \alpha\mathbf{a}^* + \beta\mathbf{c}^*$. This suggests that one should use a four-dimensional reciprocal lattice in this case (I know that this is difficult to envisage, but we are talking mathematics here!). Diffraction spots for which $m = 0$ are the main reflections; however α and β were found to be temperature-dependent, and so they were generally irrational numbers. This meant that this material did not have pure lattice periodicity. We call such a material *incommensurately modulated*. I recall de Wolff reporting on this idea at the International Congress of Crystallography in Kyoto in 1972, and that it was met by considerable disbelief, even hostility, by the audience. But we now know of many examples of incommensurate crystal structures, and so de Wolff's interpretation is fully accepted these days. In a further development, the Dutch physicists, Aloysio Janner (1928–) and Ted Janssen (1936–), showed how incommensurate structures could be described by using the symmetries of space groups in four dimensions.

We can gain a simplified understanding of these types of structures, using the idea of convolution to understand the diffraction effects from such crystals. To do this, we can denote the incommensurate structure C_{inc} as follows:

$$C_{inc} = M * [L \times O],$$

As before, L is a regular lattice, and M is a group of atoms or a molecule. The function O is some sort of modulation function which distorts the periodicity of the lattice. The Fourier Transform, using the convolution theorem, is then given by

$$C_{inc}(FT) = M(FT) \times [L(FT) * O(FT)].$$

Suppose, for example, O is a sine wave, causing the lattice to be distorted sinusoidally. The Fourier Transform O(FT) of a sine function is simply given by sharp spikes either side of the origin of the Fourier Transform. The distance between these spikes is proportional to the reciprocal of the sine period. Convolution with the reciprocal lattice then gives rise to satellite spots either side of the main spots. If these extra spots lie on rational multiples of the underlying reciprocal spacing, then one has a crystal with a commensurate *superstructure*. On the other hand, if they lie at irrational fractions of the underlying reciprocal lattice, in real space there is in a long-period sinusoidal modulation of the crystal structure that is not commensurate with the unit cell repeat.

Quasicrystals

Just when you think that everything about crystal symmetry must be known, nature retaliates by throwing a curveball. We have already seen that the basic idea of a periodic lattice is not always true in real crystals, but the scientific world was unprepared for a major discovery which subsequently altered our perception of what is meant by a crystal.

In 1984, electron diffraction patterns of a rapidly cooled metallic Al_4Mn alloy were observed by the Israeli-born materials scientist, Dan Shechtman (1941–), to show sharp spots arranged in a 10-fold symmetry, a phenomenon that has since been seen in many other complex alloys. This seemed to show that these materials in some way violated the rule that crystal lattices (with long-range translational symmetry) could not show 5-fold symmetry (10-fold symmetry is observed because diffraction patterns are always effectively centrosymmetric, thus turning 5-fold into apparent 10-fold axes).

Such materials were termed *quasiperiodic crystals* or *quasicrystals*, and they clearly could not be explained by conventional crystal symmetry ideas. Originally, this discovery was met with disbelief, and Shechtman's original paper was rejected for publication; even the Nobel laureate Linus Pauling claimed that the effect could be explained as a form of multiple twinning, i.e. by superposition of the diffraction patterns of crystals in different orientations. However, Shechtman's electron diffraction work was extremely precise and methodical, and to his credit he pursued this discovery against all the opposition. He was subsequently proved correct and was eventually awarded the 2011 Nobel Prize in Chemistry for this discovery.

We now know that such quasicrystals are not particularly rare, especially in metal alloys, and one can even grow crystals with observable faces arranged in 5-fold symmetries. For example, the alloy $Al_{63}Cu_{24}Fe_{13}$ can be grown as a faceted, single quasicrystal up to $1\,mm^3$ in size. As a consequence of Shechtman's discovery, in 1992 the IUCr revised the definition of a crystal to mean 'any solid having an essentially discrete diffraction diagram'. I have to confess that I do not know what the word 'essentially' means in this definition, and indeed defining a crystal in this way is debatable in my view.

How then can we explain this apparent violation of basic crystal lattice symmetry? A convenient, although not the only, way to do this is to think of different ways in which to tile a floor without leaving gaps. In the conventional approach, if we use a set of identical regular tiles, one ends up with periodicity and the restriction that 5- and 7-fold symmetries are not allowed. However, if one is prepared to use differently shaped tiles, then one can obtain many arrangements with local 5- or 7-fold symmetries. Such arrangements are not periodic in the traditional sense, but nonetheless create arrangements that are ordered in that they occur according to certain rules.

This has an old history in fact. As far back as 1619, Kepler showed how to fill a two-dimensional space with *different* 5-fold symmetric tiles, and in 1981 Alan Lindsay Mackay (1926–) showed

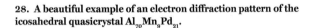

28. A beautiful example of an electron diffraction pattern of the icosahedral quasicrystal $Al_{70}Mn_9Pd_{21}$.

how to obtain a 10-fold symmetric diffraction pattern using a non-periodic pattern of points.

Figure 28 shows an example of an electron diffraction pattern of a quasicrystal of the metal alloy $Al_{70}Mn_9Pd_{21}$. This remarkable picture clearly shows 10-fold symmetry. Also note that, unlike the diffraction pattern from a normal crystal, the spots in any row are not periodically spaced but adopt a more complex set of spacings. Figure 29 shows an example of tiling in two dimensions where two different shapes are used, from an idea by Roger Penrose (1931–) in Oxford. We see that with appropriate rules for stacking, fat and thin rhombs can be placed together to fill two-dimensional space. The thin rhomb has internal angles of θ (= 36°) and 4θ,

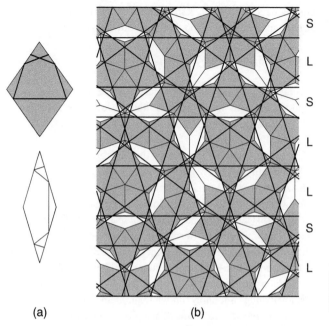

29(a). Special lines marked on the fat and thin rhombs of a Penrose tiling; (b). the effect of these special lines is to generate a Fibonacci sequence of spacings in five directions at 72° to one another.

while the fat rhomb has internal angles of 2θ and 3θ, which allows for the filling of space because 10θ = 360°. Such an arrangement never repeats, although it does show local orientational symmetry of a 5-fold kind.

Interestingly, if one superimposes a series of lines, due to Robert Ammann (1946–1994), as shown in Figure 29(b), we find that the sets of lines together show 5-fold symmetry! Furthermore, the lines are spaced by long (L) and short (S) distances in the sequence

L S L L S L S L…etc.

This sequence has an interesting property. If you replace every L by L S and every S by L, you find the following sequence

L S L L S L S L L S L L S ... etc.

This is simply a copy of the original sequence. This type of sequence is found in many places in nature and was discovered by the Italian mathematician Leonardo Pisano Bigollo (*c*.1170–*c*.1250), also called Leonardo of Pisa, or simply Fibonacci. It is interesting to note that the ratio L/S = τ, is the 'golden mean' $\tau = (1+\sqrt{5})/2 = 1.61803\ldots$ etc., well known to artists who use this to design a pleasing layout for a painting. Thus we see that this arrangement is not random, nor is it periodic. Instead, the term *quasiperiodic* is used to describe it. A note on terminology: the term *aperiodic* applies to those cases where long-range periodicity is not a feature of the solid, such as in amorphous (non-crystalline) materials like glass. The term *aperiodic crystal* is used to describe crystals where translational periodicity does not apply but where, nonetheless, sharp diffraction maxima are found. Thus incommensurate crystals and quasicrystals can be called aperiodic crystals, but not amorphous materials.

The subject becomes even more intriguing since quasiperiodicity can be shown to arise by considering a regular lattice of points in a six-dimensional space. (I know that is not easy to visualize, but mathematicians have no problem with contemplating it.) If a cut is then made through such a lattice and projected on to two- or three-dimensional space, obviously the projection will consist of points. If the cut and projection are made in the right way, it turns out that an array of points can be obtained with 5-fold symmetry. In addition, the positional ordering of the points along particular directions is again found to obey the Fibonacci sequence. This sequence is observed also in the series of spots in the diffraction patterns.

The Penrose tiling model was derived according to the strict rules of the Fibonacci sequence, in which each separation determines the next (once started, the exact sequence is derived without error), and so it is a perfect example of exact, but quasiperiodic, two-point correlation (i.e. it is a fully ordered structure). Since this is the only type of correlation present, the diffraction pattern is expected to consist of sharp spots alone.

In retrospect, perhaps, crystallographers should have known better. After all, when one thinks about how a crystal grows, it is unlikely that the depositing molecules somehow know about the long-range periodicity that we normally associate with crystal structures. Instead, each molecule, when joining together with others, only knows about its immediate surroundings. Interestingly, several years ago it was observed that when crystals of sodium chloride started to grow in solution flashes of light occurred, a phenomenon called *triboluminescence*. One of the possible explanations for this was that the initial local accumulation of atoms forms some sort of local random cluster, but that as it increases in volume a build up of strain occurs. This eventually causes the cluster to flip suddenly into an ordered arrangement, in this case with long-range periodicity, thus releasing a flash of energy in the form of light. It is also interesting to note that it can be shown, for example by the work of the Russian mathematician Boris Nikolaevich Delaunay (or Delone) (1890–1980), that one can create a lattice based solely on considering the local symmetry of points.

There is now a vast literature on the mathematics and experiments related to quasicrystals, yet one of the areas that has not been completely solved so far is the location of the atoms in the quasicrystal. The solution of quasicrystal structures is a challenge for the future. I think that crystallographers owe a debt to Shechtman for forcing us to think more generally about crystal symmetry.

Disorder

Ever since the first X-ray diffraction experiments were carried out in 1912–13, weak continuous diffuse scattering has been observed between the Bragg reflections. Laue showed in his first paper that occupational disorder in a mixed crystal gives rise to a diffuse component of the diffraction pattern. It was soon realized that such scattering is a general symptom of inherent disorder of some kind in the crystal structure. In practice not all crystal structures are perfectly ordered (in fact, they never really are!). Instead, they have various amounts and kinds of disorder. This arises whenever there is a departure from perfect ordering of the atoms in the crystal. For instance, the thermal vibration of atoms represents disorder about the mean positions of the atoms.

Similarly, disorder occurs if there is some substitution of certain atoms by different atoms, or if, say, the orientation of molecules varies in a non-repetitive array throughout the crystal. When X-rays, neutrons, or electrons are diffracted by such imperfect materials, extra scattering effects can often be observed. In addition to the appearance of the usual sharp Bragg reflections, disorder adds extra scattering into the background. This is known generally as *diffuse scattering*. Therefore the intensity of diffraction can be written as the sum of two terms:

$$I_{total} = I_{Bragg} + I_{diffuse}.$$

By conservation of energy, the more disorder that is present the lower I_{Bragg} and the higher $I_{diffuse}$. Now, if we measure the Bragg intensities only, this gives information about the *average* crystal structure, whereas the diffuse intensity tells us about the *departures* from the average structure.

To understand this further, consider for simplicity a one-dimensional crystal with atoms or molecules denoted by the letter A. Suppose that the structure looks like this:

Example 1
AAAAAAAAAAAAAAAAAAAAAAAAAAAAAAAAAA
AAA...etc.

This is obviously a perfectly long-range ordered crystal, and so, in this case, there will be no diffuse intensity. This is of course an idealized situation, because we cannot ignore the fact that the atoms or molecules must exhibit some thermal motion, even at the lowest temperatures, thus giving rise to some background diffuse scattering. Now suppose some of the As are replaced by an alternative atom or molecule, which I shall denote by the letter B. This could, for example, be by a simple substitution, or perhaps by a molecule like A but in a different orientation. Consider this structure:

Example 2
ABABABABABABABABABABABABABABABABAB
BAB...etc.

You can see that this too is a fully ordered structure with alternating As and Bs. Again, we do not expect any diffuse scattering apart from thermal scattering. Note too, in this case, that the unit cell repeat, given by ABA, is double that in Example 1. The result is that in addition to the Bragg reflections seen for Example 1, extra peaks will occur halfway in reciprocal space, thus forming superstructure reflections.

We now introduce some elements of disorder into these structures. For instance, consider this:

Example 3
AAAAAAAAAABBBBAAAAABBBBBBBBBAAAAA
BBBBBB...etc.

This structure is clearly not fully ordered but instead shows elements of *short-range order*. Every now and again, after a sequence of As a mistake occurs, and we get a few Bs. And then some As, and so on. The lengths of these correlated regions vary statistically. The effect of such a structure is to build up diffuse intensity concentrated around the Bragg reflections.

In the following we see correlated regions with alternating As and Bs, but occasionally there are mistakes when two As or two Bs occur together:

Example 4
ABABABAABABABBABABABABABBABABABA
BAABAB...etc.

This too gives rise to diffuse intensity, but this time it will be concentrated around the superstructure spots corresponding to the doubling of the basic unit cell repeat. This particular example of short-range order is an example of what is called *stacking-fault* disorder.

From this discussion, it should be clear that, whenever there is some degree of short-range order in a crystal, we should expect to see some diffuse scattering in addition to the usual Bragg reflections. This may appear as streaks or general blobs distributed in different parts of the diffraction pattern. The shorter the range of correlation the broader the diffuse features will appear, and so by studying the extent and form of the extended diffuse scattering in the diffraction pattern one can gain useful information about the nature of the disordering present. This can be useful, for instance, in helping to understand the properties of the material in question, since it is the breaking of symmetry by disordering that often explains why a particular material functions the way it does.

An interesting example is shown by the structure of the pharmaceutical aspirin. This is known to crystallize in two polymorphic forms, labelled Form I and Form II. Figure 30 shows

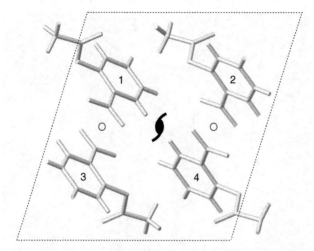

30. Crystal structure of aspirin Form II. Four complete molecules are shown.

the crystal structure of Form II. Four molecules of acetyl salicylic acid are contained within the unit cell. Molecules 1 and 3 form a dimer, i.e. two molecules linked together, related by a centre of symmetry (the small circle). Similarly, molecules 2 and 4 together are also a dimer, and the two dimers are symmetry-related by a 2-fold screw axis perpendicular to the plane of the drawing (the conventional symbol for a screw axis is shown in the centre of the unit cell).

Figure 31a shows an X-ray photograph, taken with a long exposure, revealing evidence of vertical streaks in addition to the normal Bragg reflections. The inset on the right shows a particular region.

To explain this, the authors of this work considered the four molecules 1, 2, 3, and 4 as a single unit denoted A. This unit can exist in a different orientation, denoted B. Figure 31b (left) shows a small portion of a model used to simulate the ordering: this is

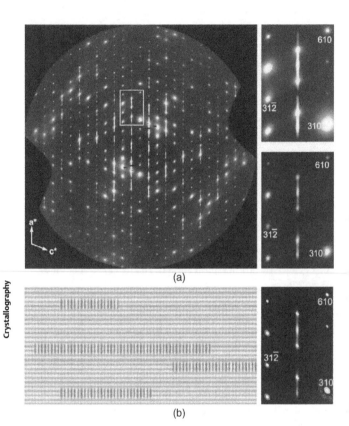

31(a). X-ray diffraction pattern for Form II aspirin (portion of the pattern is magnified on the right); (b). simulation of diffuse scattering.

represented by light and dark shapes. The A units (light grey) are highly ordered in both directions. However, units in orientation B occasionally occur (dark grey) that are more highly correlated in the horizontal direction than vertically. The longer-range horizontal order causes the diffuse scattering to be narrow in the horizontal direction, whereas the shorter-range vertical order results in the scattering being extended in the vertical direction. It can be seen that the model fits the observed diffuse scattering rather well (inset on the right). Such studies are important in understanding the properties of a material. In something like aspirin this disorder is likely to affect the solubility and hence the speed with which the pharmaceutical might be absorbed.

Diffraction

Chapter 5
Seeing atoms

The phase problem

Why don't crystallographers use X-rays to image atoms directly? To answer this question, consider first of all how an image is formed with visible light and a lens. Figure 32 shows a ray diagram for a simple lens setup with an object (here the Eiffel Tower) that is to be imaged. The rays enter the lens L and then pass through the focal plane F of the lens as shown, ending up at I, where an inverted and magnified image is obtained. Notice that on going from O to I each ray at the bottom of the object ends up at the equivalent place in the image, and similarly for the top and obviously for all points in-between. However, at any point on the focal plane information is received simultaneously from *all* points in the object. This is where the diffraction information corresponding to the object is found. The lens serves to combine all the waves from the object using their amplitudes and phases.

But what if we do not have a lens? In this case, the amplitudes and phases cannot be combined directly, and so we have to accept only the diffracted intensity information. Note that in the electron microscope there are magnetic lenses, so that it is possible simply by pressing a button to use a subsidiary lens to look at the focal plane (diffraction information) or at the image plane (where

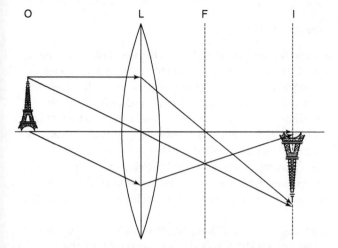

32. Formation of an image by a thin lens. O: object, L: lens, F: focal plane, I: image plane.

individual atoms and molecules can be observed). Great progress has been made in recent times to enable the electron microscope to see individual atoms. But, there are no convenient lenses suitable for X-ray and neutron beams, and so the phase information is lost.

Instead, the crystallographer measures the intensity of reflection from each plane *hkl* and from this obtains the modulus of the amplitude, called the *structure factor* $F(hkl)$ thus

$$|F(hkl)|^2 \propto I(hkl).$$

However, in order to reconstruct the crystal structure from which the diffraction information has come, it is necessary to do what the lens does for visible light. This means that the different relative phases have to be found somehow. This is known as the *phase problem*. To do this, crystallographers have developed a number of different mathematical methods to find the phases.

'Seeing' atoms

Suppose for now that we do have all the amplitudes and phases. Then the different amplitudes and phases must be combined together to form an image of the crystal structure. In the old days this was done by hand, and as several hundred (or these days thousands) of reflections had to be measured, it was incredibly time-consuming. Modern computers have changed all that. Thirty or so years ago, a graduate student in small-molecule crystallography could gain a PhD by solving typically three crystal structures containing up to one hundred atoms each. This is not the case today, where structure solution has become more routine.

The technique for combining amplitudes and phases was first suggested by WHB in his Bakerian Lecture of 1915. His technique is known as Fourier synthesis. In essence, it is a process of back-Fourier Transforming the list of structure factors together with their phases to calculate the electron density (in the case of X-ray diffraction) or nuclear density (in neutron diffraction). The electron density at any position x,y,z in the unit cell is given by

$$\rho(xyz) = \frac{1}{V}\sum_{hkl}\left|F(hkl)\right|\cos 2\pi\left(hx + ky + lz - \phi_{hkl}\right),$$

where V is the volume of the unit cell. This formula may look complicated at first, but in fact it simply represents the way in which all the diffracted waves are added together taking into account their orientations, amplitudes, and relative phases. The cosine term is just a mathematical way to describe each wave, and the phase angle ϕ_{hkl} determines where the peaks and troughs of a wave occur. Thus, if the phase angle is zero, the crest of the cosine wave is at the unit cell origin ($x = y = z = 0$): if the phase angle is 180° then a trough lies at the unit cell origin. The height of the wave, in other words its amplitude, is given by $|F(hkl)|$. The Miller indices h, k, and l determine the orientations of the waves.

This will become clearer by reference to Figure 33, which illustrates the process in two dimensions. In the left-hand diagram in Figure 33a two reflection spots, with $h = 2$, $k = -3$, and its centrosymmetric opposite $h = -2$, $k = 3$, are shown with a reciprocal lattice vector drawn between them through the origin at the centre. The middle diagram shows a set of fringes corresponding to a plane wave (as expressed by the cosine function in the equation for the electron density) perpendicular to this vector. The wavelength of the wave corresponds to the interplanar d-spacing for the particular (hkl) plane, and is inversely related to the length of the reciprocal lattice vector. The amplitude of the wave $F(2\,\overline{3}\,0) = F(\overline{2}\,3\,0) = 43$ determines how high the crests of the wave are, in this case indicated by the amount of blackening. The right-hand diagram shows this wave again. Suppose now we add a contribution from another reflection (b), in this case $h = 3$, $k = 4$ and its opposite.

The right-hand diagram now shows the result of adding this to the first wave. Notice that both waves have a phase angle equal to 0°, and this means that we place the crest of the wave through the origin, here at the bottom left. In (c) a third wave with $h = 0$, $k = 2$ is added, this time with a phase equal to 180°, placing a trough at the origin. Already with just three waves you can begin to see possible atomic positions. In (d) I show the result after adding a further 124 waves, and now the atoms making up a molecule (the carbon atoms in naphthalene in this case) can clearly be seen. The ripples around each atomic image result from what is known as *series termination*, caused by having only a finite number of reflection amplitudes—the more reflections, the less series termination and the sharper the images of the individual atoms.

Figure 34 shows a Fourier map of a molecule of naphthalene made as long ago as 1949 using an early computer to sum up 612 structure factors. One has to admire the stamina of the early

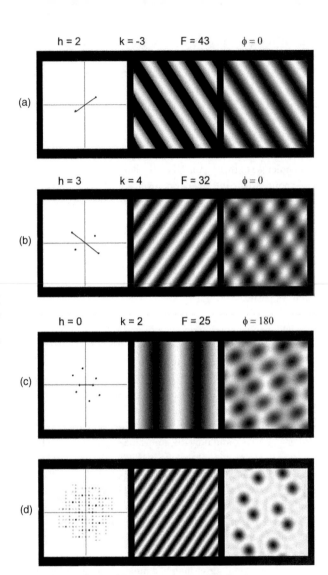

33. Fourier synthesis of the structure of naphthalene.

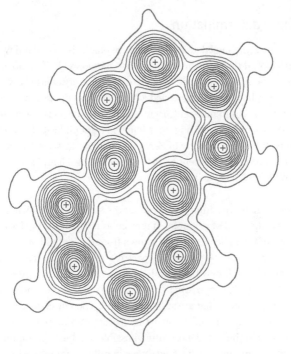

34. An early computer calculation of the electron density of a molecule of naphthalene.

researchers who had to measure the intensities of all these reflections painstakingly by eye. These days the intensities are measured automatically by diffractometers, and computers enable us to sum thousands of structure factors routinely. Fourier synthesis therefore enables us to 'see' actual atoms. In fact, it is this addition of amplitudes with phases that enables us to see objects around us: the eye lens effectively performs a Fourier synthesis for us. In a sense, therefore, in crystallography 'the crystallographer is the lens'!

Phase determination

It is clear that knowledge of the relative phases is essential if we wish to find the atoms in a crystal. So what do we do if we do not have phase information? For centrosymmetric crystals only two types of phase angle are needed, 0° and 180°, as in our example in Figure 33; but in non-centrosymmetric cases the phases can take any value and so the situation is much more complicated. Over the years, since the discovery of X-ray crystallography scientists have developed many different ways to attack this problem. I shall describe just a few of them here.

Often much prior information about the structure is known from the beginning. For example, in molecular compounds the basic shape of the molecule may be known from the method of chemical preparation, or from spectroscopy and other techniques. In such cases, it may be possible to use a trial and error method by starting with approximate positions of some of the atoms and leaving out those that are still to be found. Then by comparing the observed values of the structure factors with those calculated with this limited model it is often possible to adjust the atomic positions and include missing atoms in order to minimize the differences.

One way to do this is to repeat the Fourier synthesis, but this time using the *differences* between observed and calculated structure factors for the initial model, a so-called difference Fourier map. This often suggests likely positions for the missing atoms. In the case of small-molecule crystallography, computer programs allow the atoms to be shifted slightly by a technique known as least-squares minimization. Refinement of the atomic positions is carried out in order to reduce the differences between observed and calculated data. It is often the case that refinement by least-squares can also lead to information about the atomic vibration amplitudes as well, since the relevant displacement parameters can be included as part of the refinement.

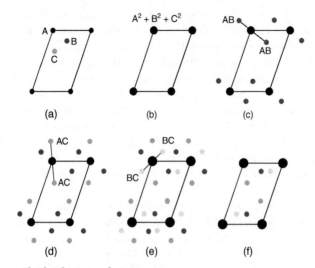

(a)   (b)   (c)

$A^2 + B^2 + C^2$

AB   AB

(d)   AC   AC   (e)   BC   BC   (f)

35. The development of a Patterson map.

Another approach is the Patterson method, due to Arthur Lindo Patterson (1902–1966) from the US in 1935. This is an important technique for obtaining a model structure without knowing the phases. A Fourier map is constructed using the intensities (to be strictly correct, the squares of the moduli of the structure factors) with all contributions given the *same* phase of 0°. The resulting map contains peaks that represent *vectors* between atoms (rather than atoms themselves), and a skilled crystallographer can use this information to suggest a structural model that fits this set of vectors. To understand this, bear in mind the phrase 'all vectors to a common origin', a favourite saying by the crystallographer C. H. (Harry) Carlisle (1911–1995) when he taught this subject at Birkbeck College London. He used to call this 'the peasant's definition of a Patterson'.

In Figure 35, I illustrate the relationship between the Patterson map and the crystal structure. Figure 35a is the projection of a hypothetical structure consisting of three atoms whose electron densities are A > B > C, respectively. Now, consider a cell of the same size as the unit cell and on it add peaks corresponding to vectors between the atoms. Thus in (b) each atom has a vector to itself, which is of zero length and so creates a peak of height $A^2 + B^2 + C^2$ at the origin of the cell (and also by translational symmetry at the other corners). In (c), peaks corresponding to vectors from A to B have been added as well as from B to A, each vector being drawn from the cell origin. The heights of these peaks are proportional to the product AB of the densities. In (d) this is repeated for vectors A to C and C to A, and in (e) from B to C and C to B; (f) shows the Patterson peaks just within a unit cell.

It can be seen that if we started with the Patterson map (Figure 35f) it should, in principle, be possible to suggest likely models for the crystal structure that would give rise to such a set of peaks, and this is where, in the old days, the skill of the crystallographer was important. The Patterson maps gain in complexity as more atoms become present in the unit cell. This method has been used for many structure determinations, for example, in the determination of the structure of benzoyl penicillin by Dorothy Hodgkin in 1945, and in the first protein structure determinations of myoglobin and haemoglobin by Max Perutz and John Kendrew in the 1940s. Today, however, less skill is required, as computers can solve this problem automatically and the Patterson solution is part of the normal package of programmes used by crystallographers. Perhaps this has taken away some of the fun in solving Patterson maps.

Suppose that in a crystal there is an atom that scatters much more strongly than the others. In this case, the other atoms can be ignored in the first instance, and the structure factors and phases can be calculated on the basis of having a single atom in the unit cell. These phases can then be applied to the observed structure factors, and through Fourier Transformation an electron density map can be computed. As the heavy scatterer is so dominant, most of the phases should be approximately correct and the map should then reveal the presence of most of the remaining light atoms. This is of particular use in macromolecular crystallography.

A related technique is to grow a crystal in which a particular strongly scattering atom is added, typically by soaking the native crystal in a solution containing the heavy atom. This causes small, but discernible, changes in the intensities of the reflections, and by using the Patterson method the light atoms can often be found. This is known as multiple isomorphous replacement (MIR), and requires at least three crystals.

The wavelength tunability of synchrotron radiation provides another method, multiple-wavelength anomalous dispersion (MAD), which is often used in macromolecular crystallography. By tuning the wavelength so that the heavier atom in the structure strongly absorbs the incident X-rays Friedel's Law is broken. Typically, measurements are made at three wavelengths, two either side and one at the absorption peak. The variation in diffracted intensities then enables the phases of the reflections to be determined.

Perhaps the most common method used in protein crystallography is *molecular replacement*. This relies on knowing the structure of a related protein, or even the same protein known from a different crystal form. To build up an atomic model of the new crystal form, it is necessary to work out how the model should be oriented and positioned in the new unit cell. This can be achieved by comparing the relative Patterson maps. Molecular replacement can be used to solve a structure when there is a good model for typically 25–30 per cent sequence similarity of the structure in the crystal. As the database of solved structures gets larger and larger, this method becomes ever more useful.

A completely different approach is to adopt more statistical approaches, known as direct methods. We already know that the amplitudes in the diffraction pattern arise from a product of the

Fourier Transforms of the atoms in a unit cell and the reciprocal lattice, so that it is obvious that there must be some relationship between the phases of various reflections. Furthermore, we know that the electron density is always positive.

An important contribution was made in 1952 by the American crystallographer David Sayre (1924–2012), who introduced the *Sayre equation*. This is a mathematical relationship that enables probable values for the phases of some diffracted beams to be found. To do this, we choose any two structure factors with indices hkl and $h'k'l'$. Then

$$F(hkl) = \sum_{h'k'l'} F(h'k'l')\, F(h-h', k-k', l-l'),$$

which implies that the structure factor for the hkl reflection can be calculated from the sum of the products of pairs of structure factors whose indices sum to the desired values of h, k, and l. One consequence of this, at least for centrosymmetric structures, is that this equation leads to the so-called triplet relationship

$$S(hkl) \approx S(h'k'l')S(h-h', k-k', l-l'),$$

where the S is the sign of the structure factor: if the phase is 0 then S is positive, negative if the phase is 180°. The relationship holds best for strong reflections. Jerry Karle (1918–2013) and Herbert Hauptman (1917–2011), both from the US, were awarded the Nobel Prize in Chemistry in 1985 for developing ways of utilizing direct methods like this to solve the phase problem. There are now several computer programs that have been developed to make use of direct methods. They work well, especially for small-molecule structures that contain many atoms of comparable scattering power, such as in organic compounds. In such cases, the determination of the structure is virtually automatic, although I should emphasize that structure determination still does throw up surprises, and it is, therefore, unwise to rely unquestionably on the automatic solution offered.

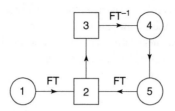

36. The charge-flipping algorithm.

In recent years, a wonderful new technique for solving crystal structures has appeared, known as *charge flipping*. This was introduced into crystallography by two Hungarian optical physicists, Gábor Oszlányi and András Sütő, in 2004 using a mathematical trick that had been employed earlier for image processing of photographs. It can be used to solve the structures of inorganic and organic crystals, where a relatively small number of atoms are to be located.

This proceeds as follows (Figure 36):

1 Start by giving *random* phases to all the structure factors.
1 → 2 Fourier Transform these structure factors and random phases to form a pseudo-electron density map. At this stage, of course, the map bears no resemblance to the true electron density map.
2 → 3 Now here's the clever trick. Take all electron density values below some assumed level and reverse their signs.
3 → 4 Back-Fourier Transform this map to generate a new set of structure factors and phases.
4 → 5 Apply these new phases to the original structure factors.
5 → 2 Fourier Transform again to obtain the next pseudo-electron density map.

This procedure is now cycled continuously, and the magical thing about it is that no symmetry information is required, and yet eventually the correct crystal structure automatically emerges as if from a mist. This trick has been applied to many crystal structure determinations and is now one of the standard routines written into crystal refinement software. I do recommend that you try out the Java applet for simulating charge flipping, listed in the Further Reading at the end of this book.

Chapter 6
Sources of radiation

X-rays

To observe diffraction from crystals it is necessary to have a source of radiation whose wavelength is of the same order as the atomic spacings. Fortunately the X-ray region of the electromagnetic spectrum happens to do this nicely. The production of X-rays in Roentgen's time and for several years after was a relatively crude affair, involving the collision of a beam of electrons with a metal target, or anticathode. Early X-ray apparatus was dangerous, leaky, and unreliable, and required control of the gas inside a glass tube, and a good vacuum in tubes manufactured later. Apparatus for producing X-rays has greatly improved since then, gaining in intensity and ease of use.

Modern reliable X-ray tubes can be found in research laboratories throughout the world (Figure 37). Electrons are emitted from a heated tungsten filament and are driven by a high-tension voltage, typically between 10 and 100 kV, towards a metal target. X-rays are then produced during the process of slowing down and absorption of the electrons by the target metal and are emitted through beryllium windows. The target is usually cooled by water to avoid overheating and damage to the target material, although in recent designs several commercial air-cooled sources are available. Higher intensity can be obtained by using a

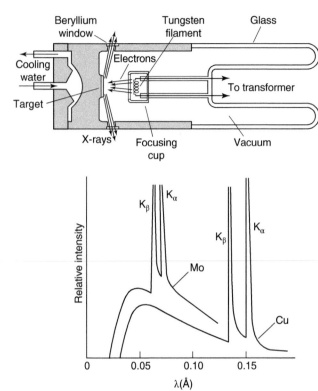

37. Schematic cut-away view of a modern sealed X-ray tube and typical emission spectra for copper and molybdenum targets.

rotating-anode generator, where cooling is achieved by using a motor to rotate the anode at high speed so that heating is not allowed to build up at one spot.

The spectrum of X-rays produced consists of intense, sharp—so-called characteristic—lines that are emitted when outer-shell electrons fill a vacancy in the inner shell of atoms in the metal target, releasing X-rays in a pattern that is 'characteristic' of each element. Characteristic X-rays were discovered by Charles Glover

Barkla (1877–1944) in 1909, who won the 1917 Nobel Prize in Physics for his discovery.

X-ray tubes

The usual lines that are used in crystallographic applications are the Kα (this is actually a close doublet). For instance, with a copper target the doublet is at $K\alpha_1 = 1.54051$ Å and $K\alpha_2 = 1.54434$ Å) and occasionally Kβ at a shorter wavelength. If one wants to have an approximately monochromatic Kα beam, one can remove most of the Kβ radiation using a filter made of a suitable metal. For example, if the target is made of copper, then a nickel filter absorbs a great deal of the emitted radiation whose wavelengths lie below the Kα wavelength. For higher-wavelength resolution, a crystal, such as silicon or graphite, is used as a monochromator by orienting a particular crystallographic plane in the X-ray beam to obey Bragg's Law. The diffracted beam will then have the particular wavelength required.

Note that in addition to the characteristic spectrum, there is a broad background radiation continuously distributed over all wavelengths. This continuous background is called white radiation or *Bremsstrahlung*. It has been well known since the 19th century that when a charged particle undergoes an acceleration or deceleration, electromagnetic radiation is emitted, the so-called Larmor radiation. This is used, for instance, in the production of radio waves. *Bremsstrahlung* arises as a result of the incident electron being decelerated on approaching atoms in the target, and it is this radiation that luckily enabled Friedrich and Knipping to obtain those first diffraction spots in the spring of 1912.

Synchrotron radiation

Early studies of the light emitted from the Crab nebula showed that it was strongly polarized, especially in the blue region. In

1953, it was suggested that this was caused by electrons following curved paths travelling at near relativistic velocities. This is known as *synchrotron radiation*.

When a charged particle is made to accelerate to near the speed of light, it can be shown that this radiation is confined to within a narrow cone that precedes the particle itself. The instantaneous power radiated by a charged particle accelerated in a circular orbit is inversely proportional to the fourth power of its rest mass, and so the radiation emitted is significant only for electron accelerators. In order to do this on earth, synchrotrons have been built in which electrons are made to travel around a circular orbit, several metres across. The radiation emitted covers a continuum of wavelengths that peaks at a wavelength proportional to the radius of the orbit and inversely as the cube power of the electron energy (typically in the GeV region). Originally, before about 1970, such machines were mainly built for high-energy physics research, where there was a need to create fast-moving particles, and the radiation was largely

Table 2

CANDLE	Armenia
Australian Synchrotron	Australia
Laboratório Nacional de Luz Sincrotron	Brazil
Canadian Light Source	Canada
Beijing Synchrotron Radiation Facility	China
National Synchrotron Radiation Laboratory	China
SSRF—Shanghai Synchrotron Radiation Facility	China
Institute for Storage Ring Facilities	Denmark
European Synchrotron Radiation Facility	France
SOLEIL	France
Angströmquelle Karlsruhe—ANKA	Germany
BESSY II—Helmholtz-Zentrum Berlin	Germany
Dortmund Electron Storage Ring Facility	Germany

ELSA—Electron Stretcher Accelerator	Germany
Metrology Light Source	Germany
PETRA III at DESY	Germany
Centre for Advanced Technology	India
Iranian Light Source Facility	Iran
DAFNE	Italy
Elettra Synchrotron Light Laboratory	Italy
Aichi Synchrotron Radiation Centre	Japan
Hiroshima Synchrotron Radiation Centre	Japan
Photon Factory	Japan
Ritsumeikan University SR Centre	Japan
Saga Light Source	Japan
SPring-8	Japan
Ultraviolet Synchrotron Orbital Radiation Facility	Japan
SESAME	Jordan
Pohang Light Source	Korea
Dubna Electron Synchrotron	Russia
Kurchatov Synchrotron Radiation Source	Russia
Siberian Synchrotron Research Centre	Russia
TNK	Russia
Singapore Synchrotron Light Source	Singapore
ALBA	Spain
MAX IV Laboratory	Sweden
Swiss Light Source	Switzerland
National Synchrotron Radiation Research Centre	Taiwan
Synchrotron Light Research Institute	Thailand
Diamond Light Source	UK
Advanced Light Source	US
Advanced Photon Source	US
Centre for Advanced Microstructures and Devices	US

Continued

Table 2 Continued

Cornell High Energy Synchrotron Source	US
National Synchrotron Light Source II	US
Stanford Synchrotron Radiation Lightsource	US
Synchrotron Ultraviolet Radiation Facility	US

ignored. However, since then the radiation has become the prime area of interest and now there are a huge number of synchrotron sources (Table 2) in the world dedicated to producing radiation, representing a multi-billion-dollar international investment.

The basic design of synchrotrons involves creating electrons with a hot-wire source and then accelerating them in a linear accelerator, essentially a long pipe surrounded by high-power magnets to steer and accelerate the electrons. The electrons emerging from the linear accelerator are then injected into a circular ring where strategically placed magnets force the electrons around a curved path. In early synchrotrons this injection of electrons into the ring took place continuously, but later this was replaced by storage rings, where the electrons were injected only occasionally (every few hours) and the electrons circulated continuously around the ring, gradually losing energy until the next injection. Many third-generation sources operate in top-up mode, whereby the loss of electron current is compensated for by relatively frequent injections. This not only provides a constant source intensity, it also means that the thermal loading across the synchrotron is constant, improving the positional stability of the beam at each beamline.

A typical storage ring layout is shown in Figure 38(a). Electrons are created at (1), linearly accelerated along the path (2), and injected into a booster synchrotron (3), which, in the case of the Diamond Light Source in the UK, uses a radio-frequency voltage source to accelerate the electrons from 100 MeV up to 3 GeV. These high-energy electrons are then injected into the storage ring (4). Since the X-radiation precedes the electrons as they travel around

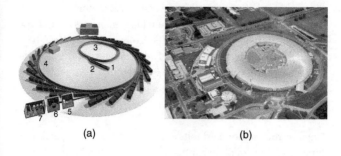

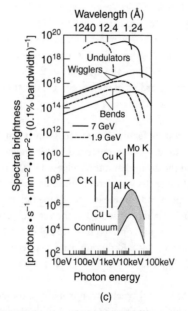

(c)

38. **Synchrotron radiation: (a). schematic view of construction; (b). aerial view of the Diamond Light Source in the UK; (c). graph of spectral brightness for different X-ray sources.**

the ring, beam pipes are built at several places around the ring tangential to the ring itself. The beam pipes then pass the radiation into specially built stations where the X-ray experiments can be carried out. Each station consists of an optics hutch (5) where the beam is focussed with a chosen wavelength, the experimental area (6), and a control cabin (7). Figure 38(b) shows a general view of the Diamond Light Source. This ring is 562 metres in circumference.

Synchrotron radiation exhibits a number of special properties. First of all, the radiation emitted by a modern synchrotron source ranges from the hard X-ray region, through the ultraviolet and infrared wavelengths, up to visible light. Unlike the X-ray tube, there are no characteristic lines; instead the intensity distribution with wavelength is continuous, so such sources are white (Figure 38c). In addition, the intensity of the X-radiation produced is many orders of magnitude greater than from a conventional tube.

Another interesting feature is that the X-ray beam, unlike in a conventional tube, is plane-polarized within the horizontal plane (viewed from the ground and thus edge-on to the ring, the electron motion as it sweeps round the ring resembles a giant horizontal dipole and this creates the plane polarization). Finally, the radiation is highly collimated in the vertical plane, e.g. at a wavelength of 1 Å it is confined to within about 0.25 milliradians (approximately 0.014°). Radiation from a conventional X-ray tube is strongly divergent.

In addition to normal synchrotron machines, sometimes extra magnets are placed in the ring to cause the electrons to execute momentarily a wiggling motion in their path. This means that locally the electrons follow a smaller diameter orbit, and so the 'wiggler' shortens the peak wavelength (i.e. it increases the peak energy) emitted towards the hard X-ray region. By inserting into the ring a periodic array of dipole magnets (known as an undulator), so that the static magnetic field alternates along the length of the periodic array, the emitted radiation from each magnet is made to sum together coherently to produce an even higher intensity concentrated within a narrow energy band in the spectrum. The undulator can be tuned to produce higher harmonics and thus get other wavelengths.

Use of synchrotrons does incur the inconvenience that one has to apply for beam time and then travel to the synchrotron for the experiments, in contrast to the ease with which one can gain access to a conventional source in the home laboratory. But it is clear that synchrotron radiation has very different properties from that produced by a conventional X-ray tube, and this means that one can carry out very different experiments from those in a normal X-ray laboratory. Probably the greatest use of synchrotron radiation these days is in the life sciences where rapid collection of large amounts of data is required in order to solve the structures of thousands of crystals of viruses and proteins.

Free-electron laser

A more recent development has been the construction of free-electron lasers. Here pulses of electrons are made to accelerate along a linear path where magnets shape the pulses to make them extremely short. These pulses then pass through an undulator, and as the electrons wiggle up and down they create radiation which in turn interacts with the electrons to create a high-intensity coherent and tuneable X-ray beam (I recommend that you view the excellent video explanation made by SLAC in the US, see Further Reading). There are now a number of free-electron laser centres in existence, with several more planned. Table 3 shows a list of the present centres.

The resulting radiation has recently been demonstrated to enable extremely rapid diffraction data collections on very small crystals of proteins. An important new development is illustrated in Figure 39. In experiments carried out in Hamburg and Stanford by

Table 3

Centre Laser Infrarouge d'Orsay	France
European XFEL	Germany
FLASH at DESY	Germany
Free-Electron Laser at ELBE	Germany
FERMI	Italy
IR FEL Research Centre	Japan
SPring-8 Angstrom Compact Free-Electron Laser	Japan
Swiss Free-Electron Laser	Switzerland
Free-Electron Laser for Infrared experiments	The Netherlands
TARLA Infrared FEL and Bremsstrahlung Facility	Turkey
Institute for Terahertz Science and Technology	US
Jefferson Lab FEL	US
Linac Coherent Light Source	US

Henry Chapman (1967–), Janos Hajdu (1948–), and John Spence (1946–), a fine powder of tiny protein crystallites is injected through the beam of X-rays from a free-electron laser. The energy of the X-rays is such as to destroy each tiny crystal, but not before a diffraction pattern is recorded on an imaging detector. Each flash image can be recorded within approximately twenty-five femtoseconds (a femtosecond is 10^{-15} seconds), and many hundreds or thousands of diffraction images are collected. As the crystallites pass through the beam, the diffraction patterns are recorded from all their random orientations. Post-processing by computer has enabled these to be sorted out, and the data used to solve the protein crystal structures. As this technique works with such tiny samples, it means that it is not necessary to have to grow single crystals on the scale used in synchrotron crystallography, and it is

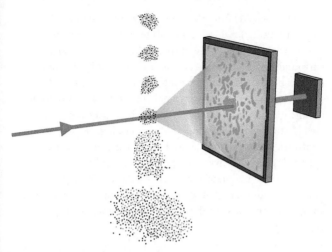

39. Use of free-electron laser radiation in protein crystallography.

predicted that in the foreseeable future it will be possible to carry out diffraction studies of single molecules, thus obviating the need for crystals at all, at least in the case of large macromolecules.

At the present time, the main disadvantage lies in the need for large quantities of protein nanocrystals (and thus purified protein). In addition, the free-electron laser provides only a single source of X-rays, thus limiting the number of users who can be accommodated at any one time. It remains to be seen if this problem can be overcome eventually. One thing is for sure: this is a fast-developing and exciting new field of research, and we can expect huge advances in this type of crystallographic research in the future.

Neutron sources

It was discovered in the mid-1940s that beams of neutrons could be diffracted by crystals in much the same way as X-rays. Bertram Brockhouse (1918–2003) from Canada and Clifford

Shull (1915–2001) from the US were awarded the 1994 Nobel Prize in Physics for developing this technique. The main feature is that, whereas X-rays are diffracted by the cloud of electrons surrounding the atoms, neutrons are scattered by the much smaller nuclei at the atomic centres. Neutrons, being subatomic particles, are themselves very small, of comparable size to the atomic nuclei. Traditionally, they have been produced by steady-state nuclear reactors that produce high-energy neutrons. These are then slowed down using moderating materials such as heavy water or graphite until their energies lie in the thermal region of about 0.12 eV. Although they are particles, under the 1924 wave–particle duality principle of de Broglie they can also be described by a wave whose wavelength is given by

$$\lambda = \frac{h}{mv},$$

where h is Planck's constant, m is the neutron mass, and v its kinetic velocity. Neutrons in the thermal regime have velocities of a few kilometres per second, and correspond to having a kinetic energy of a fraction of an electron volt. An energy of 0.12 eV corresponds to a wavelength of approximately 1.5 Å, similar to that of Cu Kα X-rays.

As the atomic nucleus is much smaller than the cloud of electrons around the atoms, neutrons are scattered over a much wider angle by the nucleus than X-rays are by the electron distribution (recall the Fourier Transform 'rules'). The neutron scattering factor, or neutron scattering length as it is usually known, therefore shows virtually no angle dependence, unlike in the case of X-rays. In addition, the magnitude of the neutron scattering lengths is not related to atomic number, and can even be negative (signifying a change in phase by 180°).

An alternative source of neutrons is offered today by the so-called spallation sources, such as ISIS at the Rutherford-Appleton

Laboratory in the UK and IPNS at the Argonne National Laboratory in the US (a European Spallation Source is currently under construction in Lund, Sweden). Here, a synchrotron is employed to accelerate protons up to high energies, after which the protons are ejected to collide with a metal target such as tungsten, thus producing neutrons. Such sources are pulsed, because of the time structure of the proton synchrotron, and span a continuum of wavelengths. They are considerably more intense sources of neutrons than is achievable from steady-state reactors, and the method of measuring the diffraction information is quite different. In this case, the time of flight of the neutrons is measured, and this gives a measure of the neutron momentum, and hence its energy or wavelength. Measurements are made using detectors set at fixed angles, and are generally not made to scan through varying angles as is normal with detectors at steady-state reactors. The data are binned into channels according to the time of flight. A plot of intensity against time of flight is then obtained. Table 4 is a list of neutron diffraction facilities worldwide.

Neutron sources have many useful applications, and the results can be complementary to those obtained from X-rays. In particular, because the X-ray scattering factors depend on the number of electrons, it can be difficult to use X-rays to locate light atoms when in the presence of heavy atoms. On the other hand, since the neutron scattering lengths are not related to the mass of the atoms directly this difficulty is overcome. However, locating hydrogen atoms using neutrons can be problematic since the hydrogen atom has a mass close to that of the neutron: the result is that when the hydrogen atom is struck by a neutron it absorbs some of the neutron energy, and this creates a large amount of incoherent background scattering. In order to get around this, sometimes the hydrogens in a crystal are replaced by the heavier isotope deuterium.

Another important use of neutron diffraction is in the study of magnetic structures. The neutron is a particle with nuclear spin,

Table 4

Bragg Institute, ANSTO	Australia
Canadian Neutron Beam Centre, Chalk River	Canada
Institut Laue–Langevin, Grenoble	France
Leon Brillouin Laboratory, Saclay	France
Berlin Neutron Scattering Centre	Germany
GEMS at Helmholtz-Zentrum Geesthacht	Germany
Juelich Centre for Neutron Science	Germany
FRM-II, Munich	Germany
Budapest Neutron Centre	Hungary
Bhabha Atomic Research Centre, Mumbai	India
ISSP Neutron Scattering Laboratory, Tokai	Japan
JAEA Research Reactors, Tokai	Japan
KENS Neutron Scattering Facility, Tsukuba	Japan
Hi-Flux Advanced Neutron Application Reactor	Korea
RID, Delft	Netherlands
Frank Laboratory of Neutron Physics, Dubna	Russia
St Petersburg Neutron Physics Institute, Gatchina	Russia
SINQ, Paul Scherrer Institut (PSI)	Switzerland
ISIS-Rutherford-Appleton Laboratory	UK
Oak Ridge Neutron Facilities (SNS/HFIR)	US
Los Alamos Neutron Science Centre (LANSCE)	US
University of Missouri Research Reactor Centre	US
Indiana University Cyclotron Facility	US

and this means that it can interact with a magnetic field. If in the crystal there are atoms that possess a net magnetic moment they interact with the neutron spin to produce extra scattering related to the locations of the atomic magnetic moments.

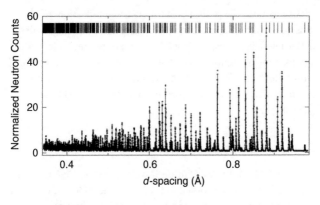

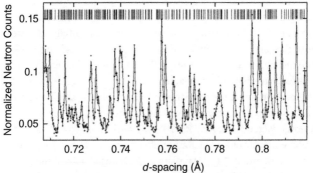

40. **Rietveld refinements using the HRPD diffractometer at ISIS in 1987. Top: aluminium oxide Al_2O_3; bottom: benzene C_6H_6.**

One of the most widely used aspects of neutron diffraction in crystallography is in the field of powder diffraction. By using the Rietveld process it is possible, especially with spallation sources, to obtain extremely high-resolution powder data and then refine the structures of the crystals making up the powder. A couple of examples are shown in Figure 40. The top diagram shows a plot of the observed data (points) together with a fit (continuous line) obtained after refining by the Rietveld program. The vertical lines at the top mark the positions of the expected reflections, and illustrate just how complex this powder pattern is. The powder pattern here is plotted against d-spacing; sometimes it is plotted against time of flight instead. The lower plot is part of the refined pattern for benzene at low temperature, and here one can see just how nicely the Rietveld process has managed to fit the peak shapes.

Electron diffraction

Between 1921 and 1925, Clinton Davisson (1881–1958) and Lester Germer (1896–1971) worked in the US on the effects of firing an electron beam from a heated filament on to a polycrystalline sample of nickel under a vacuum. As so often happens in the history of great scientific advances, air was accidentally admitted into the vacuum chamber, and this produced an oxide film on the nickel. They tried to remove this by heating the specimen, not realizing that, in so doing, the nickel formed large single-crystal areas. On redoing their experiment, they picked up a diffraction pattern on a moveable detector. At around the same time, George Paget Thomson (1892–1975) working in Aberdeen independently obtained similar effects with electrons incident on metals, and in 1937 Davisson and Thomson shared the Nobel Prize in Physics for demonstrating that electrons showed wave-like behaviour, in agreement with de Broglie's hypothesis of wave–particle duality.

Unlike with X-rays and neutrons, beams of electrons, being charged particles, can be displaced by a magnetic field. In Germany, Hans Busch (1884–1973) showed that this could be used to build an electromagnetic lens in 1926. This led to the development of the electron microscope in 1931 by the physicist Ernst Ruska (1906–1988) and the electrical engineer Max Knoll (1897–1969), both in Germany. Ruska was awarded the 1986 Nobel Prize in Physics for this work. It was soon realized that by using an electron microscope one could choose either to image a specimen or to examine its diffraction pattern.

The electron microscope, unlike with X-ray and neutron equipment, allows one to observe either the diffraction plane or the image plane, by the simple switching in or out of a magnetic lens. This provides a very powerful tool for studying the microscopic structure of materials, as one can then relate features seen in the diffraction pattern to those seen in the image. Since the original development of the electron microscope, a host of different techniques have evolved. For example, TEM is transmission electron microscopy, where the beam passes through the specimen. SEM is scanning electron microscopy, where the beam is scanned across the specimen and is reflected back into a detector. According to the different excitation potentials used to accelerate the electrons, one can study different regimes. Thus LEED, or low-energy electron diffraction, is used to study surface structures.

In recent times, enormous advances have been made in lens and detector resolution and stability, so that now it is becoming quite commonplace actually to image the individual atoms in a crystal. This is of great importance in understanding the details of crystal structures, including any distortions or disorder that may affect the material's properties. Because of heating by the electron beam and the need for a high vacuum (pressures less than 10^{-12} atmospheres), studying organic crystals, and especially biological samples, has always been a problem. However, cryogenic techniques have been installed into many electron microscopes to stabilize these samples in the electron beam, and this approach shows considerable promise.

A recent TEM development has been in a technique known as annular bright field (ABF) imaging. Here, a beam of electrons (Figure 41(a)) is focussed to sub-Ångstrom dimensions, and then scanned across the specimen. The scattered beams are detected by

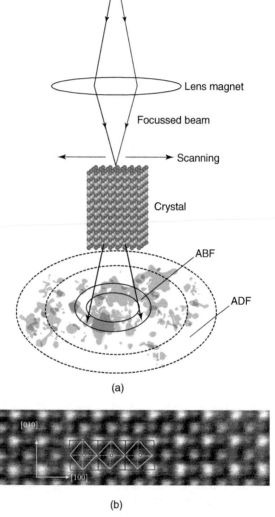

(a)

(b)

41(a). Annular bright and dark field imaging; (b). ABF image of a crystal of PbZr$_{0.53}$Ti$_{0.47}$O$_3$.

an annular detector which integrates the scattered intensity during the scanning process. The result is an image of the specimen in which individual atoms can be seen. Figure 41(b) shows an example with a specimen of a piezoelectric crystal of PZT. The lines indicate the oxygen atoms forming octahedra surrounding the Zr/Ti atoms. The Pb atoms are the large spheres in-between the octahedra, which are described by the positions of the small spheres.

Another variant is annular dark field (ADF) imaging, which collects its data from much higher angles, mainly from incoherently scattered electrons, as opposed to Bragg-scattered electrons. This is highly sensitive to variations in the atomic number of atoms in the sample. This technique is also known as high-angle annular dark field imaging (HAADF). It is clear that with the latest developments in electronics, computers, and magnet designs, there is a fascinating future for the electron microscopy of crystalline materials.

Outlook

> When men speak of the future, the gods laugh.
>
> Chinese Proverb

Crystallography, as we have seen, is an old subject, initially mainly of academic interest to scientists. In 1912 that changed, when the discoveries of Laue and the Braggs made it a science with important practical uses. Since then it has been in a state of continuous development, making major impacts on science and society through its underpinning of so much that we take for granted today. And yet, despite over a hundred years of structural studies, it is clear that major advances are still being made today. This can be seen in the creation of radical new designs for apparatus such as diffractometers, making them ever faster and more automated. Then there is the amazing development in the last forty years of new sources of radiation, using laboratory X-ray generators, synchrotrons, free-electron lasers, and neutron facilities. At the same time computing power has increased dramatically, and taking all this together we see that today scientific investigations using crystals can be made that were impossible to imagine not long ago.

Developments in magnetic lens design in electron microscopy are also leading to closer examination of the atoms in a crystal and an ability to study their departures from regular periodicity. New detectors for the electron microscope, such as used in a new technique called Cryo-EM, are also showing great promise in the study of protein structures. The free-electron laser is fast becoming a major radiation source of interest, and will certainly lead to new ways of studying solid-state structure. It is possible that in the not too distant future it may not be necessary always to have crystals, especially in the case of some biological molecules, where diffraction from individual molecules may be possible very soon. As I write, imaging of an individual virus particle has been achieved.

Interestingly, in a recent blog the protein crystallographer Gregory Petsko has written: 'The most definitive statement we can make about the future of X-ray crystallography is that it has no future in its present form!' However, in answer to this I would say that he is seeing this from the narrow point of view of a protein crystallographer, where it may well eventually be true in many cases. One needs to recall that for the protein or virus crystallographers the interest is in the molecule itself rather than how the molecules pack together in a crystal. For most, if not all, of them the crystal is simply a means of placing protein molecules in a regular array so as to obtain a diffraction pattern. They care little for the reasons for the particular packing arrangements and the interactions between the protein molecules in a crystal. But X-ray crystallography (and neutron and electron crystallography) is much more than the study of proteins, important though they are. However, I do agree that it will not remain *in its present form*, as this is a subject that is changing rapidly.

One of the areas that is set to grow is the use of solid-state nuclear magnetic resonance (NMR) to determine crystal structures. Improvements in the design of NMR instruments are beginning to produce interesting information complementary to the usual diffraction methods.

New, useful solids are being discovered all the time, and often make the news. We saw this with the discovery several years ago of high-temperature superconductors, which spawned numerous scientific publications and an enormous amount of public interest. A recent material that is causing a stir is that of graphene and its related substances. Even more recently it has been discovered that highly efficient solar cells can be made using the class of materials called perovskites. These are currently hot topics, and huge numbers of publications are appearing on them as I write. These materials all need crystallographic methods to be applied in order to understand them, modify them, and adapt them for practical usage. Some of them may change our lives dramatically. No doubt more wonder materials will be found in the years to come. I suspect that in a hundred years' time Bragg's Law will continue to be widely used.

Further reading

Authier, André. 2013. *Early Days of X-ray Crystallography*. Oxford University Press.

Bernstein, Joel. 2008. *Polymorphism in Molecular Crystals*. Oxford Science Publications on behalf of the International Union of Crystallography.

Blow, David. 2002. *Outline of Crystallography for Biologists*. Oxford University Press.

Branded, Carl-Ivar and John Tooze. 1998. *Introduction to Protein Structure*. Taylor and Francis.

Brown, Andrew. 2007. *J. D. Bernal: The Sage of Science*. Oxford University Press.

Burke, John G. 1966. *Origins of the Science of Crystals*. University of California Press.

Burns, G. and A. M. Glazer. 2014. *Space Groups for Solid State Scientists*. Academic (Elsevier).

Clegg, William. 2015. *X-ray Crystallography*. Oxford University Press.

Ewald, P. P. Editor. 1962. *Fifty Years of X-ray Diffraction*. N. V. A. Oosthoek's Uitgeversmaatschappij for the International Union of Crystallography.

Ferry, Georgina. 1998. *Dorothy Hodgkin: A Life*. Granta.

Ferry, Georgina. 2008. *Max Perutz and the Secret of Life*. Pimlico.

Glazer, A. M. 1987. *The Structures of Crystals*. Adam Hilger.

Glazer, Mike, and Patience Thomson. 2015. *Crystal Clear: The Autobiographies of Sir Lawrence and Lady Bragg*. Oxford University Press.

Glynn, Jennifer. 2012. *My Sister Rosalind Franklin*. Oxford University Press.

Hall, Kersten T. 2014. *The Man in the Monkey Nut Coat: William Astbury and the Forgotten Road to the Double Helix*. Oxford University Press.

Hammond, Christopher. 2015. *The Basics of Crystallography and Diffraction*. Oxford Science Publications on behalf of the International Union of Crystallography.

Hargittai, István. 1992. *Fivefold Symmetry*. World Scientific.

Hargittai, István, and Magdalena Hargittai. 1994. *Symmetry: A Unifying Concept*. Shelter.

Hunter, Graeme. 2004. *Light is a Messenger: The Life and Science of William Lawrence Bragg*. Oxford University Press.

Jenkin, John. 2008. *William and Lawrence Bragg, Father and Son: The Most Extraordinary Collaboration in Science*. Oxford University Press.

Lima-de-Faria, J. 1990. *Historical Atlas of Crystallography*. Springer.

Maddox, Brenda. 2003. *Rosalind Franklin: The Dark Lady of DNA*. Harper Collins.

McPherson, Alexander. 2009. *Introduction to Macromolecular Crystallography*. Wiley-Blackwell.

Senechal, Marjorie. 2013. *Shaping Space: Exploring Polyhedra in Nature, Art, and the Geometrical Imagination*. Springer.

Watson, James D. 2012. *The Double Helix*. Simon and Schuster.

Crystallography on the Internet

There are a huge number of articles, videos, and courses on Crystallography available on the Internet. I list a few to get you started. You can find many more by searching with Google, but do not be tempted by the Crystal Healers!

Attar, Naomi. 2013. 'Raymond Gosling: The Man who Crystallized Genes'. <http://genomebiology.com/2013/14/4/402>.

Boyle, P. D. 'Growing Crystals that will Make Your Crystallographer Happy'. <http://xray.chem.uwo.ca/crystal_growing/GrowXtal.html>.

Bragg, Melvyn. 2014. 'Bragg on the Braggs'. <http://www.bbc.co.uk/programmes/b0383vb0>.

Bragg, Melvyn. 2013. 'In Our Time: Crystallography'. <http://www.bbc.co.uk/programmes/b01p0s9s>.

Charge flipping. A Java applet demonstrating how charge-flipping works. <http://escher.epfl.ch/flip/>.

Cryostream. <http://www.oxfordcryosystems.com>.

Curry, S. 2013. 'The Science Behind the New Foot-and-Mouth Disease Vaccine'. <http://www.theguardian.com/science/occams-corner/2013/mar/30/1>.

Diamond Light Source. A film dealing with the historical development of modern crystallography. <http://www.richannel.org/the-braggs-legacy>.

e-Crystallography course. <http://escher.epfl.ch/eCrystallography/>.

Glazer, A. M. Videos about crystallography. <http://www.amg122.com/twobraggs/public.html>.

Glazer, A. M. Videos of famous crystallographers. <http://www.amg122.com/twobraggs/videos.html>.

Glazer, A. M. Demonstration of Fourier synthesis. <http://amg122.com/programs/Fourier.html>.

Introduction to Crystallography (MIT Course). <http://tinyurl.com/pou994h>.

Nuffield Department of Medicine. 2014. 'History of Structural Biology'. <http://www.ndm.ox.ac.uk/part-2-the-history-of-structural-biology>.

Nugent, Keith A. 2015. 'Viewpoint: X-Ray Imaging of a Single Virus in 3D'. *Physics*, 2 March. Imaging of a single virus particle using the Free Electron Laser <http://physics.aps.org/articles/v8/19>.

Petsko, G. 'Crystallography without Crystals'. <http://cen.xraycrystals.org/essay-on-the-future-of-crystallography.html>.

Protein crystallization. <http://en.wikipedia.org/wiki/Protein_crystallization>.

Royal Institution Crystallography Collection. Many excellent videos on crystallography. <http://www.richannel.org/collections/2013/crystallography>.

Royal Society of Chemistry website on synchrotron radiation. <http://www.rsc.org/education/eic/issues/2011May/DiamondLightSource.asp>.

Service, Robert F. 'Electron Microscopes Close to Imaging Individual Atoms'. A description of Cryo-EM. <http://news.sciencemag.org/biology/2015/05/electron-microscopes-close-imaging-individual-atoms>.

SLAC YouTube video on the Free Electron laser. <http://www.youtube.com/watch?v=zKbJMFFjnNU>.

Crystal drawing programs

Crystal Maker. Crystal and Molecular Structures Modelling and Diffraction. <http://www.crystalmaker.com/> (commercial).

Crystallographica Toolkit. <http://www.oxcryo.com/cg/crystallographica/> (free).

DIAMOND (Crystal and Molecular Structure Visualization). <http://www.crystalimpact.com/diamond/> (commercial).

PYMOL (Protein drawing). <http://www.pymol.org/> (free).

Shape <http://www.shapesoftware.com/00_Website_Homepage/> (commercial).

SMORPH crystal shapes. <http://www.smorf.nl/draw.php> (free).

VESTA (Visualization for Electronic and Structural Analysis). http://jp-minerals.org/vesta/en/ (free).

WinXMorph crystal shapes. <http://cad4.cpac.washington.edu/WinXMorphHome/WinXMorph.htm> (free, donations requested).

Crystallographic databases

CCDC (Cambridge Crystallographic Data Centre—Organic Crystals). <https://www.ccdc.cam.ac.uk/pages/Home.aspx>.

Crystallography Open Database. <http://www.crystallography.net/>.

CRYSTMET (Metals and minerals). <http://www.pdb.org/pdb/home/home.do>.

ICSD (Inorganic Crystal Structures Database). <https://icsd.fiz-karlsruhe.de/search/index.xhtml>.

Pauling File (Minerals). <http://www.materialsdesign.com/medea/pauling>.

PDB (Protein Data Bank). <http://www.pdb.org/pdb/home/home.do>.

Index

A

acetyl salicylic acid 48, 91
Acharya, Ravindra
  foot-and-mouth disease virus 53
adamite 30, 31
$Al_{70}Mn_9Pd_{21}$ 84
alkali halides 18
all-face-centring 34
alpha-helix 49
Alzheimer's disease 51
amethyst xiii
amino acid 48–50
Ammann, Robert 85
amorphous 46, 86
amplitude 70–2, 74, 95–9, 100,
  103, 104
annular bright field imaging 123
annular dark field imaging 125
anomalous dispersion 71, 103
aperiodic crystal 86
Aristotle xiv
Armstrong, Henry Edward
  letter to *Nature* 19
aspirin 47, 48, 90–2, 93
Astbury, William Thomas 22
atomic position 37, 48, 100
Australian aborigines and quartz
  and amethyst xiii
average crystal structure 88

B

ball-and-spoke model 47, 48
barium titanate 53, 58
Barkla, Charles Glover
  characteristic X-rays 109
Barlow, William
  alkali halide models 19
  crystal models based on packing
    of spheres 17
  derivation of 230 space groups 10
bcc 41
benzene 47, 121
Bergman, Torbern and cleavage of
  calcite 5
Bernal, John Desmond 21
beryl xiv
beryllium 39, 107
beta sheet 50, 51
Bigollo, Leonardo Pisano
  (Fibonacci) 86
biological macromolecule 48,
  49, 51
body-centred 34, 35, 41, 53
body-centred cubic 41, 53
Bradley, Albert James 79
Bragg reflection 88–91
Bragg, William Henry 11, 15, 17, 19,
  20–2, 31, 67, 96
  Fourier synthesis 96

Bragg, William Lawrence 11,
 15–22, 31, 41, 67, 79
 sound-ranging 20
Bragg's Law 16, 17, 67, 109, 127
Branson, Herman
 alpha helix 49
Bravais, Auguste 8, 34
Bravais lattice 8, 10, 35, 37
Bremsstrahlung
 X-rays 109, 116
Bridgman-Stockbarger crystal
 growth 60
Brockhouse, Bertram
 neutron scattering 117
Broglie, Louis-Victor-Pierre-
 Raymond 17, 118, 122
Brown, Alexander Crum 19
Brown, Fred
 foot-and-mouth disease virus 53
buckminsterfullerene 45
Busch, Hans
 electromagnetic lens 122

C

caesium chloride 41
calaverite 79
calcite xiii, 5
Capeller, Maurice 1
capsid 52
Carangeot, Arnould
 contact goniometer 5
Carlisle, Charles Harold (Harry) 101
Cavendish Laboratory 21, 22
ccp 41
centre of inversion 28
centre of symmetry 28, 29, 91
centred cell 32, 34, 35
C-face centring 34
Chapman, Henry
 free electron laser 116
charge flipping 105, 106
chiral symmetry 25
chirality 9, 10, 25, 51, 72
cloud chamber 11

contact goniometer 5
Controlled Vapour Deposition 62
convolution 36, 72, 73, 75, 81, 82
convolution theorem 72, 75, 81
copper 41
copper sulfate pentahydrate 13, 14
copper target
 X-rays 109
Corey, Robert
 alpha helix 49
Cosier, John
 invention of the Cryostream 55
Crab nebula 109
Crick, Francis 21
cryocrystallography 55
cryo-EM 126
Cryostream 55, 56
crystal class 8, 10
crystal
 definition by IUCr 83
crystal growth 4, 57–8, 62
 Bridgman-Stockbarger
  growth 60
 Czochralski growth 58–60
 floating zone method 60
 flux growth 57, 58
 hydrothermal crystal growth 61
 melt growth 60
 Verneuil crystal growth 60
crystal structure xv, xvi, 2, 7, 10, 14,
  15, 17, 18, 19, 21, 30, 31, 33, 36,
  37, 39, 41–6, 48, 50, 52–54, 57,
  59, 61, 63, 69, 70, 72, 78, 79,
  81, 82, 87, 88, 91, 95, 96, 101,
  102, 105, 106, 116, 123, 127
crystal system 30, 34, 36
crystals used for rain-making xiii
cubic system 1, 14, 17, 28, 29, 31,
  34, 41, 43, 45, 53, 58
cubic close-packed structure 41
Curl, Robert Floyd 45
Cuvier, Georges 5
Czochralski, Jan
 accidental discovery of crystal
  growth method 58

D

Daniel, Vera 79
Davisson, Clinton
  electron diffraction 122
de Wolff, Pim 80
Dehlinger, U. 79
Delafosse, Gabriel
  unit cell 8
Delone, Boris Nikolaevich 87
diabetes 51
diamond xiv, 14, 19, 42–44, 56, 62
diamond anvil
  high-pressure crystallography 56
Diamond Light Source
  synchroton 111–13
difference Fourier 100
diffraction grating 12
diffractometer 18, 51, 56, 69, 70
diffuse scattering 88–90, 92, 93
dimer 91
direct methods 103, 104
disorder 48, 88–90, 93, 123
DNA 9, 21, 48, 52
double helix 9, 22

E

electron density 96, 99, 104, 105
electron diffraction 82–4, 123, 124
electron microscope 95, 96, 122,
    123, 125, 126
enzyme 22, 48, 49, 51, 52
Epicurus xiv
Ewald, Peter Paul 12, 13, 64
Ewald construction 68, 70
Ewald Sphere 66, 67

F

face-centred 17, 34, 41, 43
face-centred cubic structure 17, 41
fcc 41, 43
Fedorov, Evgraf Stepanovich
  derivation of 230 space groups 10

feldspar 21
femtosecond 116
Fibonacci sequence 85–7
floating zone method 60
fluorescence 14, 56
flux growth 57, 58
foot-and-mouth disease virus 53
Fourier map 97, 100, 101
Fourier synthesis 96, 98–100
Fourier Transform 72–5, 81, 82,
    96, 102, 104–6, 118
Fox, Graham
  foot-and-mouth disease virus 53
Frankenheim, Moritz Ludwig 8
Franklin, Rosalind Elsie 22
free-electron laser 115–17, 125, 126
Friedel's Law 71, 72, 103
Friedrich, Walter 8, 12–16, 109
Fry, Elizabeth
  foot-and-mouth disease virus 53

G

gallium arsenide 44, 60
Geim, Andre Konstantin 44
gene 48
genome 52, 53
germanium xv, 20, 43, 59
  electron diffraction 122
glide plane 37
gold 41, 79
golden mean 86
Gosling, Raymond 22
graphene 44, 127
graphite 42, 44, 109, 118
Greeks xiv, 1
group 6, 8, 9, 10, 25, 27, 29, 37, 38,
    43, 45, 50, 52, 82
Guglielmini, Domenico 5

H

haemoglobin 21, 50, 51, 102
Hajdu, Janos
  free electron laser 116

Han Xia xiv
Han Ying comparing crystals with
    flowers xiv
hanging drop vapour diffusion 63
hardness of diamond xiv, 44
Hauptman, Herbert
    direct methods 104
Haüy, Abbé René-Just 5, 6, 7, 8, 104
hcp 39
heat exchanger
    in Cryostream 55
heavy atom method 103, 119
heavy water
    nuclear reactor 118
Hessel, Johann Friedrich
        Christian 32
    crystal classes 8
hexagonal system 28, 34, 39
hexagonal close-packed
    structure 1, 39
hexamethylbenzene 21
high-pressure crystallography 56
Hodgkin, Dorothy xvi, 21, 102
Hooke, Robert 2, 3
Hume-Rothery, William 55
Huygens, Christian 2
hydrogen atoms
    location by neutron
        scattering 119
hydrogen bond 46, 50
hydrothermal crystal growth 61

Iceland spar xiii
identification of materials 78
identity operation 27, 28
incommensurate crystal 79, 80,
    81, 86
injection of electrons
    synchrotron 112
intensity of scattering 18, 71, 74,
    75, 88, 89, 90, 92, 93, 119, 123
International Notation 27, 37,
    38, 43

International Tables Commission of
    the IUCr 38
International Union of
    Crystallography 23, 26, 38, 83
ionization spectrometer 17, 18
iron crystals at centre of the
    earth xiv, xv, 54
ISIS spallation source 118, 120, 121
isomorphous replacement 103
IUCr 23, 26, 38, 83
IUCr definition of a crystal 83

J

Janner, Aloysio 81
Janssen, Ted 81
Jenkinson, C.H 17

K

Karle, Jerome
    direct methods 104
Kendrew, John 21, 102
Kepler, Johannes 1, 2, 41, 8
King's College London 2, 7
Knipping, Paul Karl Moritz
    12–16, 109
Knoll, Max
    electron microscope 122
Kroto, Harold (Harry) Walter 45

L

Larmor radiation 109
lasca 62
lattice 8–10, 32–8, 41, 43, 52,
    64–6, 70, 72–6, 78, 81–3, 86,
    87, 97, 104
lattice points 32, 34, 36, 43, 67, 75
Laue, Friedrich and Knipping's
    experiment 15
Laue, Max Theodor Felix 12–17, 21,
    40, 68, 89, 127
Laue diffraction 13, 17, 18, 67, 68
Laue equations 13

Laue photograph 13, 68
Law of Constancy of Interfacial
    Angles 4
lead zirconate 30
lead zirconate titanate (PZT) 45, 124
least-squares refinement 100
Leonardo of Pisa (Fibonacci) 86
linear accelerator 112
Lipson, Henry 79
lithium niobate xv
lock and key model 48, 49
Lonsdale, Kathleen 21, 55
low-energy electron diffraction
    (LEED) 123
low-temperature crystallography
    45, 47, 54, 56, 122
Lucretius xiv
lysozyme 22, 51

M

Mackay, Alan Lindsay 83
macromolecule 48, 51, 69, 103, 117
magnesium 39
magnetic structures
    neutron diffraction 120
measuring reflections by eye 69, 99
Megaw, Helen Dick
    Megaw Island 21
methylammonium lead iodide 45
Micrographia 2, 3
Miller index 30, 31, 70, 79, 96
Miller, William Hallowes 30
mirror furnace 61
mirror symmetry 24, 25, 37
Mitscherlich, Eilhard
    Law of Isomorphism 7
modulated crystal 79–82
molecular replacement 3
monochromator 109
multiple isomorphous replacement
    (MIR) 103
multiple-wavelength anomalous
    dispersion (MAD) 103
myoglobin 21, 102

N

naphthalene 97–9
Nazca Caves xiv
neutron diffraction 66, 88, 95, 96,
    119–21
neutron scattering length 118, 119
neutron spallation source 118, 121
Nobel Prize 10, 11, 20–2, 44, 52,
    55, 83, 104, 109, 118
    Ada Yonath 52
    Andre Konstantin Geim 44
    Bertram Brockhouse 117
    Charles Glover Barkla 109
    Clifford Shull 118
    Clinton Davisson 122
    Dan Shechtman 82, 83, 87
    Dorothy Hodgkin xvi, 21, 102
    Ernst Ruska 122
    Francis Crick 21
    George Paget Thomson 122
    Harold (Harry) Walter Kroto 45
    Herbert Hauptman 104
    James Watson 21
    Jerry Karle 104
    John Joseph Thomson 11, 17
    John Kendrew 21, 102
    Konstantin Sergeevich
        Novoselov 44
    Linus Pauling 22, 49, 83
    Maurice Wilkins 22
    Max Perutz 21, 102
    Max von Laue 12, 20
    Richard Errett Smalley 45
    Robert Floyd Curl 45
    Tom Steitz 52
    Venkatraman Ramakrishnan 52
    Wilhelm Roentgen 10–12
    William Henry Bragg 11, 14,
        17–22
    William Lawrence Bragg 11,
        15–22
non-symmorphic space group 37
Novoselov, Konstantin
    Sergeevich 44

nuclear magnetic resonance 127
nuclear reactor 118

O

optical rotation 9
optical waveguide xvi
orthorhombic system 28, 30, 52
Oszlányi, Gabor
    charge flipping algorithm 105

P

packing of spheres 3, 10, 11, 39–41
Parkinson's disease 51
Pasteur, Louis 8, 9, 25, 59
Patterson, Arthur Lindo 101
    Patterson method 101–3
Pauling, Linus 22, 49, 83
    alpha-helix 49
Peking man and quartz xiii
penicillin xvi, 21, 102
Penrose, Roger and tiling 84–7
peptide 49–51
periodicity 2, 7, 31, 33, 79–83, 86,
    87, 126
perovskite 42, 45, 127
Perutz, Max 21, 102
Petsko, Gregory 126
pharmaceutical xv, xvi, 9, 78, 90, 93
phase xv, 15, 16, 70, 71, 74,
    94–105, 118
phase problem 94, 95, 104
phase transition 53, 58
Phillips, David Chilton
    lysozyme structure 22
photograph 50
    of Rosalind Franklin 22
photovoltaic 45
piezoelectric xvi, 44–6, 60, 125
Pliny the Elder xiv
point group 27, 29
point lattice 9, 76
point symmetry 25, 32, 38
polarized light 9, 111, 116

poliovirus 53
polonium 14
polymorph 44, 78, 90
polypeptide 49, 51
Pope, William 17
powder diffraction xv, 75–9, 121
primary protein structure 50, 51
primitive lattice 32, 34, 41
primitive unit cell 32, 35
protein xv, xvi, 21, 22, 48–55, 62,
    63, 69, 70, 102, 103, 115–17, 126
protein-misfolding 51
protein structure 21, 22, 50–2, 71,
    102, 126
PZT 45, 125

Q

quartz xiii, xvi, 4, 25–7, 42–6,
    61, 62
quasicrystal 82–7
quasiperiodic crystal 82, 86, 87
quaternary protein structure 51

R

Ramakrishnan, Venkatraman
    ribosome structure 52
Réaumur, René Antoine de 2
reciprocal lattice 64–7, 70, 74, 75,
    78, 81, 82, 97, 104
reciprocal lattice vector 66, 67, 81, 97
reflection symmetry 24
ribosome structure 52, 55
Rietveld, Hugo, and Rietveld
    refinement 78, 121, 122
right-handed helix 49
RNA 52, 53
rock crystal xiii, xiv
Roentgen, Wilhelm Conrad
    10–12, 107
Romé de l'Isle, Jean Baptiste Louis
    Law of Constancy of Interfacial
        Angles 4
rotation axis 24

rotational symmetry 8, 24, 26, 32
Rowlands, David
    foot-and-mouth disease virus 53
Royal Institution 22, 55
Ruska, Ernst
    electron microscope 122

S

Saint-Hilaire, Etienne Geoffroy 7
salt 18, 19, 41, 43
sapphire and ruby crystals 60
Sayre, David
    direct methods 104
    Sayre equation 104
scanning electron microscopy 123
Schoenflies, Artur Moritz
    derivation of 230 space
        groups 9, 10
    notation 27, 30, 37
Schuster, Franz Arthur Friedrich
    theory of optics 15
screw axis 8, 10, 52, 91
secondary protein structure 50, 52
seed crystal 4, 56–62
semiconductor xv, 44, 60, 62
series termination 97
Shechtman, Dan 82, 83, 87
short-range order 90
Shull, Clifford
    neutron scattering 118
silicon xv, 20, 43, 45, 59, 69
silicon dioxide 26, 45, 46
silver 41, 79
sitting drop vapour diffusion 63
Smalley, Richard Errett 45
Smithells, Arthur 18
snow crystals xiii, 1, 2, 57
sodium carbonate 80
sodium chloride 18, 19, 43, 53, 87
Sohncke, Leonhard 8
Sommerfeld, Arnold 11–13
SONAR xvi
space group 8–10, 37, 38, 43–45,
    51, 52, 81

space lattice 8
spallation source
    neutron scattering 118, 120
stacking-fault 90
steady-state nuclear reactor 118,
    119, 120
Steitz, Tom
    ribosome structure 52
Steno, Nicolas
    Law of Constancy of Interfacial
        Angles 4
stereographic projection 26, 27
storage ring
    synchrotron 110, 112
strontium titanate 45, 60
structure factor 71, 96–106
Stuart, David
    foot-and-mouth disease virus 53
substrate 49
superstructure 82, 89, 90
surface structure 123
Sütő, Andras
    charge flipping algorithm 105
symmetry xiii, xv, 1, 4, 6, 8, 24–43,
    51, 53 70, 73–5, 79, 82–91,
    102, 105
symmetry element 8
symmetry operation 24, 25, 27,
    37, 52
symmorphic space group 37
synchrotron radiation 51–6, 70, 72,
    109–16, 119, 125

T

Tao Wang xiv
tartrates 8, 25
Teal, Gordon Kidd
    growth of silicon and germanium
        crystals 59
terrestrial magnetism 54
tertiary protein structure 50
tetragonal system 28, 30, 35, 58
thermal vibration of atoms 12, 88
Thomson, John Joseph 11, 17, 122

tiling 84–7
time of flight
    neutron diffraction 119, 122
Todd, Charles 11
translational symmetry 6, 31, 32,
    37, 73, 79, 82, 104
transmission electron
    microscopy 123
trial and error 100
triclinic system 27–29
trigonal system 28, 46
tripeptide 49
triplet relationship
    direct methods 104
tungsten target for spallation
    source 119
twinning 83

U

undulator
    synchrotron 114, 115
unit cell 6, 8, 9, 14, 17, 32, 34, 36,
    37, 41, 43, 45, 47, 48, 53, 64,
    70, 75, 79, 82, 89, 91, 96, 102,
    103, 104

V

Vegard, Lars 14
Verneuil, Auguste
    Verneuil crystal growth 60

virus xvi, 49, 52–54, 115, 128
vitamin B12 xvi, 21

W

Watson, James 21
wave-particle duality 17, 118, 122
wiggler
    synchrotron 114
Wilkins, Maurice 22
Wilson cloud chamber 11
Wilson, Charles Thomson Rees
    11, 15
women in science 21

X

Xiaodong Song xiv
X-ray crystallography xvi, 17, 48,
    60, 67, 100, 126
X-ray diffraction 39, 51, 55, 59, 67,
    79, 88, 92, 96
X-rays 11, 16, 17, 55, 64, 66, 67, 71,
    72, 74, 88, 94, 103, 107, 108,
    116–19, 122
X-rays, discovery of 10, 11, 108

Z

zinc 40
zinc sulfide 13, 14, 44
β-D,L-allose 47, 48

Expand your collection of
VERY SHORT INTRODUCTIONS

1. Classics
2. Music
3. Buddhism
4. Literary Theory
5. Hinduism
6. Psychology
7. Islam
8. Politics
9. Theology
10. Archaeology
11. Judaism
12. Sociology
13. The Koran
14. The Bible
15. Social and Cultural Anthropology
16. History
17. Roman Britain
18. The Anglo-Saxon Age
19. Medieval Britain
20. The Tudors
21. Stuart Britain
22. Eighteenth-Century Britain
23. Nineteenth-Century Britain
24. Twentieth-Century Britain
25. Heidegger
26. Ancient Philosophy
27. Socrates
28. Marx
29. Logic
30. Descartes
31. Machiavelli
32. Aristotle
33. Hume
34. Nietzsche
35. Darwin
36. The European Union
37. Gandhi
38. Augustine
39. Intelligence
40. Jung
41. Buddha
42. Paul
43. Continental Philosophy
44. Galileo
45. Freud
46. Wittgenstein
47. Indian Philosophy
48. Rousseau
49. Hegel
50. Kant
51. Cosmology
52. Drugs
53. Russian Literature
54. The French Revolution
55. Philosophy
56. Barthes
57. Animal Rights
58. Kierkegaard
59. Russell
60. William Shakespeare
61. Clausewitz
62. Schopenhauer
63. The Russian Revolution
64. Hobbes
65. World Music
66. Mathematics
67. Philosophy of Science
68. Cryptography
69. Quantum Theory
70. Spinoza
71. Choice Theory
72. Architecture
73. Poststructuralism
74. Postmodernism
75. Democracy

76. Empire
77. Fascism
78. Terrorism
79. Plato
80. Ethics
81. Emotion
82. Northern Ireland
83. Art Theory
84. Locke
85. Modern Ireland
86. Globalization
87. The Cold War
88. The History of Astronomy
89. Schizophrenia
90. The Earth
91. Engels
92. British Politics
93. Linguistics
94. The Celts
95. Ideology
96. Prehistory
97. Political Philosophy
98. Postcolonialism
99. Atheism
100. Evolution
101. Molecules
102. Art History
103. Presocratic Philosophy
104. The Elements
105. Dada and Surrealism
106. Egyptian Myth
107. Christian Art
108. Capitalism
109. Particle Physics
110. Free Will
111. Myth
112. Ancient Egypt
113. Hieroglyphs
114. Medical Ethics
115. Kafka
116. Anarchism
117. Ancient Warfare
118. Global Warming
119. Christianity
120. Modern Art
121. Consciousness
122. Foucault
123. The Spanish Civil War
124. The Marquis de Sade
125. Habermas
126. Socialism
127. Dreaming
128. Dinosaurs
129. Renaissance Art
130. Buddhist Ethics
131. Tragedy
132. Sikhism
133. The History of Time
134. Nationalism
135. The World Trade Organization
136. Design
137. The Vikings
138. Fossils
139. Journalism
140. The Crusades
141. Feminism
142. Human Evolution
143. The Dead Sea Scrolls
144. The Brain
145. Global Catastrophes
146. Contemporary Art
147. Philosophy of Law
148. The Renaissance
149. Anglicanism
150. The Roman Empire
151. Photography
152. Psychiatry
153. Existentialism
154. The First World War
155. Fundamentalism
156. Economics
157. International Migration
158. Newton
159. Chaos
160. African History

161. Racism
162. Kabbalah
163. Human Rights
164. International Relations
165. The American Presidency
166. The Great Depression and The New Deal
167. Classical Mythology
168. The New Testament as Literature
169. American Political Parties and Elections
170. Bestsellers
171. Geopolitics
172. Antisemitism
173. Game Theory
174. HIV/AIDS
175. Documentary Film
176. Modern China
177. The Quakers
178. German Literature
179. Nuclear Weapons
180. Law
181. The Old Testament
182. Galaxies
183. Mormonism
184. Religion in America
185. Geography
186. The Meaning of Life
187. Sexuality
188. Nelson Mandela
189. Science and Religion
190. Relativity
191. The History of Medicine
192. Citizenship
193. The History of Life
194. Memory
195. Autism
196. Statistics
197. Scotland
198. Catholicism
199. The United Nations
200. Free Speech
201. The Apocryphal Gospels
202. Modern Japan
203. Lincoln
204. Superconductivity
205. Nothing
206. Biography
207. The Soviet Union
208. Writing and Script
209. Communism
210. Fashion
211. Forensic Science
212. Puritanism
213. The Reformation
214. Thomas Aquinas
215. Deserts
216. The Norman Conquest
217. Biblical Archaeology
218. The Reagan Revolution
219. The Book of Mormon
220. Islamic History
221. Privacy
222. Neoliberalism
223. Progressivism
224. Epidemiology
225. Information
226. The Laws of Thermodynamics
227. Innovation
228. Witchcraft
229. The New Testament
230. French Literature
231. Film Music
232. Druids
233. German Philosophy
234. Advertising
235. Forensic Psychology
236. Modernism
237. Leadership
238. Christian Ethics
239. Tocqueville
240. Landscapes and Geomorphology
241. Spanish Literature

242. Diplomacy
243. North American Indians
244. The U.S. Congress
245. Romanticism
246. Utopianism
247. The Blues
248. Keynes
249. English Literature
250. Agnosticism
251. Aristocracy
252. Martin Luther
253. Michael Faraday
254. Planets
255. Pentecostalism
256. Humanism
257. Folk Music
258. Late Antiquity
259. Genius
260. Numbers
261. Muhammad
262. Beauty
263. Critical Theory
264. Organizations
265. Early Music
266. The Scientific Revolution
267. Cancer
268. Nuclear Power
269. Paganism
270. Risk
271. Science Fiction
272. Herodotus
273. Conscience
274. American Immigration
275. Jesus
276. Viruses
277. Protestantism
278. Derrida
279. Madness
280. Developmental Biology
281. Dictionaries
282. Global Economic History
283. Multiculturalism
284. Environmental Economics

285. The Cell
286. Ancient Greece
287. Angels
288. Children's Literature
289. The Periodic Table
290. Modern France
291. Reality
292. The Computer
293. The Animal Kingdom
294. Colonial Latin American Literature
295. Sleep
296. The Aztecs
297. The Cultural Revolution
298. Modern Latin American Literature
299. Magic
300. Film
301. The Conquistadors
302. Chinese Literature
303. Stem Cells
304. Italian Literature
305. The History of Mathematics
306. The U.S. Supreme Court
307. Plague
308. Russian History
309. Engineering
310. Probability
311. Rivers
312. Plants
313. Anaesthesia
314. The Mongols
315. The Devil
316. Objectivity
317. Magnetism
318. Anxiety
319. Australia
320. Languages
321. Magna Carta
322. Stars
323. The Antarctic
324. Radioactivity
325. Trust

326. Metaphysics
327. The Roman Republic
328. Borders
329. The Gothic
330. Robotics
331. Civil Engineering
332. The Orchestra
333. Governance
334. American History
335. Networks
336. Spirituality
337. Work
338. Martyrdom
339. Colonial America
340. Rastafari
341. Comedy
342. The Avant-Garde
343. Thought
344. The Napoleonic Wars
345. Medical Law
346. Rhetoric
347. Education
348. Mao
349. The British Constitution
350. American Politics
351. The Silk Road
352. Bacteria
353. Symmetry
354. Marine Biology
355. The British Empire
356. The Trojan War
357. Malthus
358. Climate
359. The Palestinian-Israeli Conflict
360. Happiness
361. Diaspora
362. Contemporary Fiction
363. Modern War
364. The Beats
365. Sociolinguistics
366. Food
367. Fractals
368. Management
369. International Security
370. Astrobiology
371. Causation
372. Entrepreneurship
373. Tibetan Buddhism
374. The Ancient Near East
375. American Legal History
376. Ethnomusicology
377. African Religions
378. Humour
379. Family Law
380. The Ice Age
381. Revolutions
382. Classical Literature
383. Accounting
384. Teeth
385. Physical Chemistry
386. Microeconomics
387. Landscape Architecture
388. The Eye
389. The Etruscans
390. Nutrition
391. Coral Reefs
392. Complexity
393. Alexander the Great
394. Hormones
395. Confucianism
396. American Slavery
397. African American Religion
398. God
399. Genes
400. Knowledge
401. Structural Engineering
402. Theatre
403. Ancient Egyptian Art and Architecture
404. The Middle Ages
405. Materials
406. Minerals
407. Peace
408. Iran
409. World War II

410. Child Psychology
411. Sport
412. Exploration
413. Microbiology
414. Corporate Social Responsibility
415. Love
416. Psychotherapy
417. Chemistry
418. Human Anatomy
419. The American West
420. American Political History
421. Ritual
422. American Women's History
423. Dante
424. Ancient Assyria
425. Plate Tectonics
426. Corruption
427. Pilgrimage
428. Taxation
429. Crime Fiction
430. Microscopy
431. Forests
432. Social Work
433. Infectious Disease
434. Liberalism
435. Psychoanalysis
436. The American Revolution
437. Byzantium
438. Nuclear Physics
439. Social Psychology

440. Water
441. Criminal Justice
442. Medieval Literature
443. The Enlightenment
444. Mountains
445. Philosophy in the Islamic World
446. Light
447. Algebra
448. Hermeneutics
449. International Law
450. Moons
451. Sound
452. Epicureanism
453. Black Holes
454. The Body
455. Fungi
456. The History of Chemistry
457. Environmental Politics
458. Modern Drama
459. The Mexican Revolution
460. The Founding Fathers
461. Hollywood
462. Goethe
463. Medieval Philosophy
464. Earth System Science
465. Slang
466. Computer Science
467. Shakespeare's Comedies
468. The Welfare State
469. Crystallography